COSMIC WHEEL: A JOURNEY OF PHOTONIC CONSCIOUSNESS

S. JASON CUNNINGHAM

Front-cover photo: Photograph taken by S. Jason Cunningham

S. Jason Cunningham
Cosmic Wheel: A Journey of Photonic Consciousness
www.smokeandmirrorsbook.com
Library of Congress Cataloging-in-Publication Data
Cunningham, S. Jason
Cosmic Wheel: A Journey of Photonic Consciousness / S. Jason Cunningham
p. cm.
Includes bibliographical references and glossary.
ISBN-13: 978-0-9885483-2-9
ISBN-10: 0988548321
1. Quantum physics 2. Astronomy 3. Ancient History 4. Hermeticism. 5. Science 6. Mysticism I. Title
1st Edition, May 2014

About the Author

S. Jason Cunningham has spent the last 14 years working extensively in Africa, Southwest Asia and the Middle East, specializing in conflict resolution, geopolitical and cultural research and investigations, and risk management and due diligence in such countries as Egypt, Iraq, Saudi Arabia, U.A.E., Libya, Uganda, Ethiopia, Pakistan and Afghanistan. The author's career began some 20 years ago working on a diplomatic dialogue in support of the Middle East Peace Process, and has since travelled the world visiting more than 60 countries. The author has visited ancient archeological sites throughout North and East Africa, and the Middle East and has through research and onsite work uncovered extraordinary history and ancient wisdom. S. Jason Cunningham has published two previous books, ***Smoke and Mirrors***, and ***Approaching Singularity: The Genesis of Creation***. Meet the author – www.smokeandmirrorsbook.com

DEDICATION

I lovingly dedicate this book to humanity and its return to Cosmic Consciousness.

"The lips of wisdom are closed, except to the ears of understanding."
Hermetic axiom

CONTENTS

FORWARD

Throughout the Ages, mankind has applied mind, body and spirit with great energy to grow and develop on this Earth, while starring towards the heavens and wondering about a greater force which shapes our collective destiny. Mathematicians and scientists have probed and explained, as best they could, the infinitesimally small particles of life and the infinitely expansive universe within which we exist, while philosophers and theologians have pondered the source, meaning and ultimate fate of life itself.

Scientific theories have been developed backed by advanced mathematics to explain our physical world, while broader questions regarding the essence of life and the hereafter are left to realms of spirituality. But despite the collectively enlightenment forged by mankind throughout recorded history, we really know very little beyond our 3rd dimension physical environment. And, as new theories expanding upon the human mind, body and spirit are advanced, the question which always surfaces quickly in today's society is: "Can you prove it?"

Newton's Law of Gravity connected energy with motion and then Einstein's Theory of Relativity connected energy with matter, and both created quantum leaps in our understanding of physics and the associated forces which not only shape our lives, but the very existence of our universe. But yet, science and mathematics have not been able to explain, let alone prove, the connection between the fundamental energy of the universe and the very essence of human consciousness. We know that it is light which sustains life on earth as plants, through photosynthesis, create oxygen and carbon based food. Is it possible that non-thermal light is the "missing link" between energy and our human consciousness? Can science be connected with spirituality by understanding the link between "subtle energy" and "photonic consciousness?"

Photonic Consciousness is a profound theory advanced by the author to explain the conscious, creative force of light. The case is made that biophotons are the carriers of electromagnetic radiation at varying wavelengths, and that DNA infused with biophotonic information will instruct cells accordingly. In effect, our DNA is actually "frozen light," and it is photonic consciousness that ignites our DNA and creates the energetic resonance which awakens our soul and through which we experience life itself.

By grounding this new theory in historical evidence, complex wave theory, mathematical derivations and, most importantly, photographic evidence and personal experience (which I witnessed on numerous occasions), the author presents a compelling case that there is a "cosmic consciousness" that is the field of infinite knowing and wisdom. This field can be accessed by expanding our personal "I AM" super consciousness to resonate within the harmony and frequency of the cosmic consciousness.

Previously in the book *Smoke and Mirrors* the author exposes forces which seek to trap mankind in a false matrix of reality; and, then in the book *Approaching Singularity: the Genesis of Creation* points the way for our mind and spirit to ascend through the coming cosmic singularity without fear or hesitation. In *Cosmic Wheel: A Journey of Photonic Consciousness*, your mind will be expanded beyond the known physical world of potential, thermal and kinetic energy into the new realm of "subtle energy" which supports the author's theory of "photonic consciousness." *Cosmic Wheel* reveals a new and deeper understanding of the secrets of the cosmos and the profound, energy-driven changes taking place as all of us here on Earth transit the galactic plane of the Milky Way, leaving behind the Age of Pisces and entering the Age of Aquarius.

Rear Admiral (ret) C.R. Kubic, P.E.

CHAPTER ONE

CYCLICAL SHIFTS IN CONSCIOUSNESS: THE COSMIC WHEEL OF LIFE

"You are a divine being. You matter, you count. You come from realms of unimaginable power and light, and you will return to those realms."
Terence McKenna

The Cosmic Wheel can be understood as the cyclical movement of our solar system in the Milky Way galaxy. A metaphorical cosmic wheel written about in ancient writings, mythology, and illustrated in paintings left by civilizations separated by centuries and vast geographic distances. For thousands of years, humans have told stories of the cosmic cyclical movements within time, and foretold its significance for the human experience on Earth.

As our solar system now passes through the plane of the Milky Way galaxy humanity approaches the path along the cosmic wheel wherein the Golden Age is prophesized to return.

The Golden Age is spoken of by numerous civilizations and within numerous religious and philosophical texts as a time when harmony and peace will return to Earth. The Hindu's call it "yugas" with alternating dark and golden ages. Could it be that the yugas or cyclical phases of life on Earth correspond to our solar systems movement in the Milky Way galaxy, and if so, how?

Modern science and rational thought teaches us to put aside mythology, religion and fanciful notions to seek truth through observation. In other words; if you can see it, touch it, smell it, taste it and repeat an outcome through observation and experimentation – it is real.

Within the pages of this book I will take you on a journey where science meets spirituality to prove that consciousness is both infinite and eternal, and we are a grand part of the Creator's tapestry of life within the cosmic wheel. This is the ultimate cosmic journey of Photonic Consciousness.

Milky Way Galaxy

Our Milky Way is a marvel of creation. It is a barred spiral galaxy an estimated 100,000 light-years in diameter containing 200 billion stars, and an estimated 100 billion solar systems transiting in a corkscrew manner within the galactic spiral, moving up and down through the galactic plane.

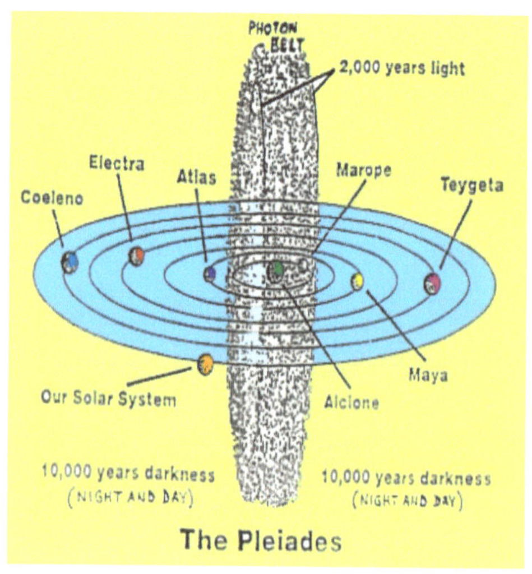

> *"Our Earth along with other planets revolve around the Sun. The Sun, along with 6 other stars, in turn revolve around a bigger star, Alcyone, also known as the Central Sun of the Pleiades. The Milky way galaxy, to which our Earth belongs, has billions of such stars and planets, all revolving around massive stars in its core, called as the Galactic Center. As our Solar System and other stars orbit Alcyone, and as Alcyone moves through the Milky Way galaxy, the star orbits (relative to the galactic center) are shaped more like spirals–or vortexes."*[1]

It is the macrocosmic orbit of stars and solar systems through the spiral of the Milky Way, to the microcosmic orbit of planets in our solar system around our sun, to the even smaller microcosmic orbit of our moon around Earth, to the even infinitesimally smaller microcosmic orbit of protons and neutrons within the atom. As the planets in our solar system orbit the sun, so to does our solar system orbit a 'central sun' located at the center of the Milky Way galaxy.

First observed in April 2007, scientists believe that at the center of the Milky Way lies an immense black hole called "Sagittarius A-Star", which resides there virtually unmoving.[2] The heart of the Milky Way is approximately 26,000 light years away from our solar system, and in our approach to the 'heart' we enter a high vibratory photon belt.

Scientists are baffled by this phenomenon and seek to understand what lies at the center of our galaxy. Scientists studying this phenomenon explain that the black hole has a

---

[1] http://prof77.wordpress.com/new-earth/crossing-the-galactic-plane-and-the-photon-

[2] http://www.space.com/15166-milky-center-black-hole-sagittariusastar.html
http://www.eventhorizontelescope.org/docs/Doeleman_event_horizon_CGT_CFP.pdf

radius no larger than about two-tenths the distance between the Earth and the Sun, and holds as much mass as four million suns.[3] It is invisible and produces much less visible light than the stars that orbit it, however the photon belt that extends from the galactic center emits powerful gamma ray radiation.

Scientists presume the mass in the galactic center is a black hole or possibly a ball of unidentified particles. They do know that objects within our galaxy are being pulled by a yet unknown force residing in the galactic center, and telescopes seeking to observe this energetic force find nothing there. This is an enigma, igniting our imagination.

In 2012, NASA has launched a project called "Event Horizon Telescope" which is a worldwide network of 50 radio telescopes that will capture a picture of the massive black hole at the center of our galaxy.[4] The name for this project is cleverly chosen, "Event Horizon", as it is the event horizon that resides at the edge of the point of singularity, which science perceives as illustrated by a black hole.

In studying the existence of black holes, scientists are using general relativity simulations to model the spectrum and structure of Sagittarius A-Star by evolving the electromagnetic and plasma fields surrounding the theorized black hole. Keep this in mind as you progress into the journey of photonic consciousness.

Knowledge of the Ancients

Much to the chagrin of mainstream science, ancient civilizations, such as the Mayans, understood centuries ago the significance of a force that resides at the galactic center they referred to as a "Central Sun".

For the Mayans the central sun, or Hunab Ku, is the

---

[3] Ibid.

[4] http://www.space.com/14278-black-hole-photos-event-horizon-telescope.html

creative force that gives "measure and motion to all things".[5] Somehow they understood that cyclical orbits do give both measure and motion to all matter and has a profound affect upon it in concurrence with its rate of vibration; meaning, high or low. They believed that it is within the pathway of the dark rift of the Hunab Ku the veil of illusion will be lifted and humanities' consciousness will ascend to birth a Golden Age – a new era of peace.

Similarly this can be understood in the teachings of Christianity as the second coming of Christ. Virtually all religions believe in the coming of a messiah, a messenger of God, who will usher in world peace. It comes as no surprise that many religions and ancient texts suggest that the golden age will last for 2,000 years. This just so happens to correspond to what modern science now understands, which is the 2,000-year transit of our solar system through the high vibratory photon belt as we move through the galactic plane. This is a cyclical process along the cosmic wheel of time that we find ourselves entering now.

Prior to immersing myself in the consciousness awakening and seeking truth in ancient wisdom, I could not understand how ancient cultures understood and prophesized such things. Were they somehow 'special', or were they able to access a cosmic information field of source energy and draw this information into their awareness?

The ancients understand a time on Earth when there existed a golden age and the gods (small "g") associated with this civilization. The gods of Sumeria, Babylonia, Egypt, Greece, Rome, Mayan, and so forth, can all be traced to a similar

---

[5] Mayan Prophecy 2012 http://mayanprophecy2012.blogspot.com/2009/07/ hunab-ku-galactic-butterfly-mayan.html

mythology that points humanity to a higher truth of its creation and a conscious energy force that governs all of existence.

What I have come to understand is that indeed there is a conscious energy force that imbues itself within a quantum information field, and within this infinite field of information resides absolute wisdom and truth. To understand the nature of this quantum information field we must look to both science and spirit. The two appear to be opposing forces, but in fact when merged into one resolve the enigma giving us a glimpse at the ineffable.

The Photon Belt

As our solar system transits through the galactic plane we enter a high vibratory space of gamma ray light that is colliding and bombarding everything at the atomic and molecular level. This is a hyper-charged photonic quantum field of energy that is radiating outward from the center of the Milky Way and is transmitting information from the quantum information field. As you will discover through the pages of this book, this quantum information field is a creative force of infinite potentiality, a force of light in which "Photonic Consciousness" travels.

> *"Our solar system goes through the belt twice each cycle of 26,000 years (that is, every half-cycle or 13,000 years). The thickness of the photon cloud is such that it takes about 2,000 years for our solar system to pass through it, and therefore about 11,000 years between each encounter with this belt. Alcyone [the central sun in the Pleiades system] is in the photon belt all the time. Like the rings around Saturn, the photon belt might be a much larger ring of energy around the star Alcyone, a ring*

parallel to the flat plane of the Milky Way galaxy. Of course, because Alcyone always remains near the center of the galactic plane, the photon belt might (alternatively) be an even larger ring or "galactic plane" of energy emanating from the Milky Way's galactic center."[6]

According to NASA, their Fermi Gamma-ray Space Telescope has unveiled a previously unseen structure located at the center of the Milky Way. The feature spans 50,000 light-years and is theorized by NASA to be the remnant of an eruption from a supersized black hole at the center of our galaxy. "What we see are two gamma-ray-emitting bubbles that extend 25,000 light-years north and south of the galactic center," said Doug Finkbeiner, an astronomer at the Harvard-Smithsonian Center for Astrophysics in Cambridge, Mass., who first recognized the feature. "We don't fully understand."[7]

GAMMA-RAY BUBBLE AT THE CENTER OF THE MILKY WAY GALAXY (Mayans referred to as the "Dark Rift")

A term I have created "Photonic Consciousness", meaning light consciousness, comes from a source I am only beginning to understand; and as I will show you with the physical

---

[6] http://prof77.wordpress.com/new-earth/crossing-the-galactic-plane-and-the-photon-belt/

[7] http://www.nasa.gov/mission_pages/GLAST/news/new-structure.html

evidence I have gathered over the last few years, this source frequency is multidimensional and originates from higher dimensional planes outside of the space-time continuum.

It is Photonic Consciousness that transmits quantum information, energy frequencies of varying levels (along the spectrum of light) into our solar system and all living matter to affect our physical and spiritual evolution. For instance, as our solar system transits closer to source energy residing in the galactic center and is immersed within the photon belt, higher frequency light is colliding at the molecular and atomic level of all matter. This collision is causing both the absorption of more light (information) and with this the ascension of the atom (matter).

I'm certain you must be thinking right now, "what?" This subject boggled my mind as I began to explore it, however the explanation as to why this happens is not as complicated as you might think.

As we learn in basic science classes, everything we perceive in this third dimensional universe is merely vibrating-oscillating atoms. Some scientists may contend that there is no conscious creation behind it that creation happens by chance. However in the field of quantum physics, some scientists are now exploring the possibility that nothing is by chance and there is a creative force governing matter.

It is very apparent to me that there is an extraordinary conscious creative power actively directing all seen and unseen forces that affect matter; and we as infinite, quantum consciousness residing in physical matter are emerging from the dense energetic bonds that have kept our consciousness in a state of separation from this truth.

As you will discover in the depths of this book, the cosmic wheel is the cyclical process built into the cosmic clock ensuring that life continues to rise onward and upward in its biological and consciousness evolution. As we enter the photon belt and take our close approach to the galactic center, photonic consciousness is streaming into our body and mind igniting our DNA and awakening our Soul to our cosmic origins as infinite, quantum Beings of Light.

CHAPTER TWO

APPROACHING SINGULARITY

"Come (you - man) from things unseen unto the end of those that are seen, and the very emanation of Thought shall reveal unto you how faith in they which are unseen was found in them which are seen, they that belong to the Unbegotten Father.
Whomsoever hath ears to hear, let him hear!"
Sophia of Jesus Christ

As I wrote in my second book, *Approaching Singularity: The Genesis of Creation*, published 2012, I was amazed to find the depth by which even ancient religious texts illustrate the great truth held within the cosmic wheel. For instance, in the Gospel of Thomas, one of the 12 Apostles of Jesus the Christ, he writes, "In the Round Dance of the Cross, Jesus asks humanity to join in the Cosmic Dance; *"Whoever dances belongs to the whole.' 'Amen.' 'Whoever does not dance does not know what happens.' Amen."*[8]

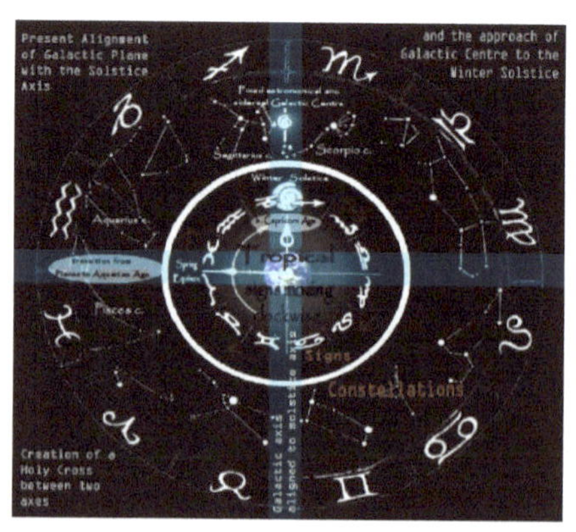

[8] Pagels, Elaine. 2004. *Beyond Belief: The Secret Gospel of Thomas*

This is a clear reference to the cyclical transit of our solar system crossing the plane of the galactic center, which is the "cross". The symbol of the cross is also a reference to the electromagnetic field that exists within the center of the Vesica Pisces, a geometric form for physical life illustrated within the golden ratio of phi, which I will discuss in a later chapter.

Some Christians may take offense to my suggestion that this is the origin of the cross, however we know that the cross has been in existence as a sacred symbol thousands of years before the birth of Jesus, and has taken on many varied forms all resembling one another. There are endless references from ancient civilizations pointing to this great truth who understood the power of God's creative force (light and sound) and the cyclical shifts of human consciousness on Earth. This truth was handed down through civilizations protected within symbols, parables, religious texts and mythology.

The Egyptian Key of Life (ankh), the Celtic Cross, the Christian Cross, the Solar Cross, Wheel Cross, the Indian Cross, the Tau Cross, the Greek Cross, the Maltese Cross, the Coptic Cross, the Chi-Rho Cross, the Rosicrucian Cross, the Alchemical sign, the mathematical sign and the astrological cross just to name a few.

Some historians who seek to disprove an extraordinary messenger of God we know as Jesus the Christ, may point to the history and iconology of the cross as a means to claim it is rooted in pagan lore, however I strongly disagree with this approach. As I present this information to you understand that I believe Jesus was a messenger of God whose divine Soul came to Earth to return this important truth to a civilization descending into the depths of darkness.

The cross does hold its origins from the Egyptian Key of Life, which symbolizes the cosmic wheel described as the

"Mirror of Venus" or "Venus Key" having the following meaning:

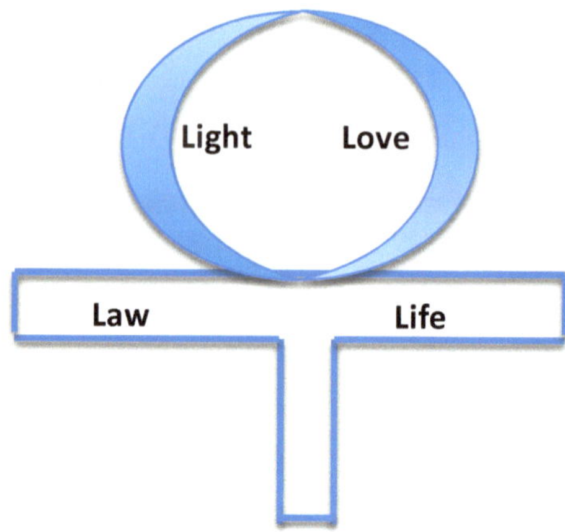

Law: The cosmic system of order that governs and balances the elements of matter.

Life: The vibratory – energetic resonance by which we experience life and the various expressions of physical existence as conveyed through the quantum information field of Photonic Consciousness.

Light: The harmonic, color-coded vibration (God's creative force) through which everything is created and resides therein.

Love: Divine illumination and the energy of perfection of self as a co-creator with the Creator God. Love is the harmonic frequency of God.[9]

Notice in the Ankh graphic above, we enter through a passage (gateway) to life and law, and ascend into the sphere of light and love.

Key of Life and the Tree of Life

Images of the Key of Life, Tree of Life, and Flower of Life all represent the same truth of the nature of creation and the

---

[9] Phylos, Orpheus. 1999. *Earth, the Cosmos and You"; Orpheus Gateways*: 23-27

quantum conscious energy that governs it. Some historians to this day believe the Egyptian Key of Life to be the symbol of a sandal, and some Egyptian historians assert that it is a representation of the Nile river, however the ankh is a powerful symbol passed down through civilizations by wise teachers to remind us of a truth we are only now re-discovering within the cosmic wheel of time.

Below is an image of the ankh and its similarity to the tree of life, human chakra system, and the sacred geometry of the flower of life.

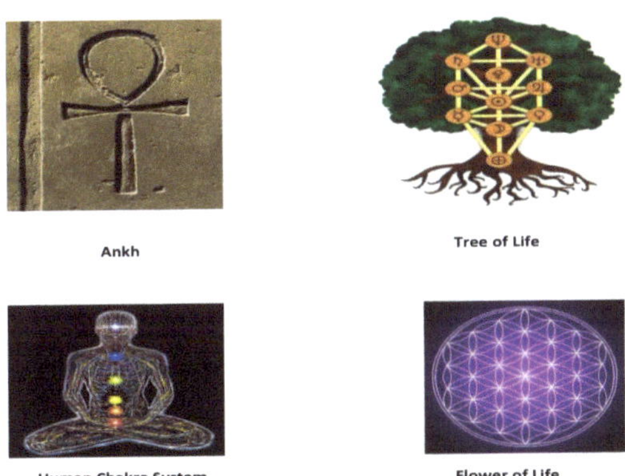

Ankh Tree of Life

Human Chakra System Flower of Life

The ancient Egyptians understood that everything is energy; everything vibrates and moves...nothing remains still. They understood the Key of Life represents aspects of creation, which correspond to their understanding of sacred geometry as reflected in the Flower of Life and Tree of Life, the energetic chakra system in the physical body corresponding to its electromagnetic field (EMF), and the Vesica Pisces representing the quantum field of the golden ratio of phi.

This information was handed down to ancient Egyptians through concepts spoken of in their mystery schools rooted in Hermeticism that spoke very clearly about the cosmic law of vibration and quantum, infinite consciousness. Below is a brief glimpse at the extraordinary truths held within the Hermetic concepts taught thousands of years ago. If you practice meditation and are familiar with the chakra energy system of the body, you may have already surmised the seven Hermetic principles align to each chakra. Below I've ordered the hermetic principles in correspondence to the charka they serve.

Seven Hermetic Principles:

7) *Principle of Mentalism: Everything is mind, the universe is mental held in the mind of the ALL (God).* [Meaning, everything is photonic consciousness. Corresponds to the crown chakra.]
6) *Principle of Correspondence: The principle of "As above, so below, as below, so above." In all things there is correspondence.* [Corresponds to the third eye chakra.]
5) *Principle of Vibration: Everything vibrates and moves, nothing is at rest.* [Corresponds to the throat chakra.]
4) *Principle of Polarity: Everything is dual and has its opposites. All paradoxes can be reconciled with the understanding that opposites are identical in nature, but merely varying in degree - rate of vibration.* [Corresponds to the heart chakra.]
3) *Principle of Rhythm: In everything there is motion, an ebb and flow, a swinging backward and forward and that rhythm compensates for this movement.* [Corresponds to the solar plexus chakra.]
2) *Principle of Causation: Every cause has its effect and everything happens according to law, nothing happens by chance.* [Corresponds to the sacral chakra.]
1) *Principle of Gender: Gender manifests in all*

dimensions and everything has its masculine and feminine principles.[10] [Corresponds to the root chakra.]

The origination of these teachings is lost to history, however the ancient Greeks spoke of a man who once resided in Egypt thousands of years before the Greek civilization giving him the name "Hermes Trismegistus" the god of wisdom, thrice great, who was known as the scribe of the gods. In ancient Egyptian hieroglyphs he is depicted holding a pen scribing the truth, or often holding the scales of justice measuring the heart of the human before passing into the afterlife.

The ancient Egyptians called him Thoth (Djehuty or Tehuti) and depicted him with the head of the Ibis or Baboon as an animal totem in reverence for the wisdom he bestowed to humanity. Modern religious scholars believe Hermes/Thoth to be the contemporary of Abraham in the Torah by which Abraham is said to have derived his wisdom. Esoteric tradition, religion, philosophy and scientific knowledge are built on the foundation of ancient hermetic teachings. Often the hermetic axiom is used to elucidate the importance of this ancient wisdom, *"The lips of wisdom are closed except to the ears of understanding."*[11]

DJED

The ancient Egyptians are the earliest known postdiluvian civilization to center its culture in the belief of the power of eternal consciousness and the cosmic influences upon it. They believed in the concept of the "Djed", which corresponds to the cosmic wheel.

The concept of the Djed has a spiritual and physical meaning. It is written as a combination of "Mer-Ka-Ba" or simply

---

[10] Three Initiates. 2011 *Kybalion: A Study of the Hermetic Philosophy of Ancient Egypt and Greece*

[11] Ibid.

"Ka or Kaw", which is the energy force of light and stability representing balance. The concept "Mer" is the light that rotates within itself, reflective of the photonic consciousness immersion in the cosmos and thus into the atom (matter).

"Ka" is the life force, and "Ba" refers to matter that conforms to the concept of reality under the creative guiding force of photonic consciousness. Thus, the "merkaba" is a cyclical, rotating light that is the portal by which consciousness ascends and descends within the multidimensional framework, a macro and microcosmic 'stargate'.

The ancient Egyptians used the sun disk as their symbol for light in all its aspects of creation, both physical and spiritual. The physical interpretation of light can be understood as energetic vibrations of the sun itself, the "solar djed", bringing both illumination and thermal heat to the physical world. The spiritual interpretation can be understood as the conscious, creative force of God, whom they referred to as "Ra", the God above lesser gods. The photonic or light iconology being that of a sun disk illustrates energy (photonic consciousness) radiating into all matter illuminating the body, mind and soul with ancient wisdom. For the ancient Egyptians, light was the ultimate creative principle and governing force of all dimensions.

As an aspect of the cosmic wheel, ancient Egyptians practiced a ceremony they called "Sa'ha Djed", meaning raising the djed or the raising of consciousness. This was a ceremony performed at the end of one world age and the birth of a new world age. The Djed is translated as, "to declare" or "to speak", and the root of the word means "culmination or azimuth of the star". The Djed represents a symbolic singularity of polarities along the cosmic wheel, and presents humanity with the opportunity for consciousness to expand and ascend with the infusion of more light.

Within the cosmic wheel symbol, the Djed marks the center axis of the galactic cross pointing to the galactic center. Extraordinarily this is the moment in which humanity now finds itself as we move into the Age of Aquarius and the photon belt.

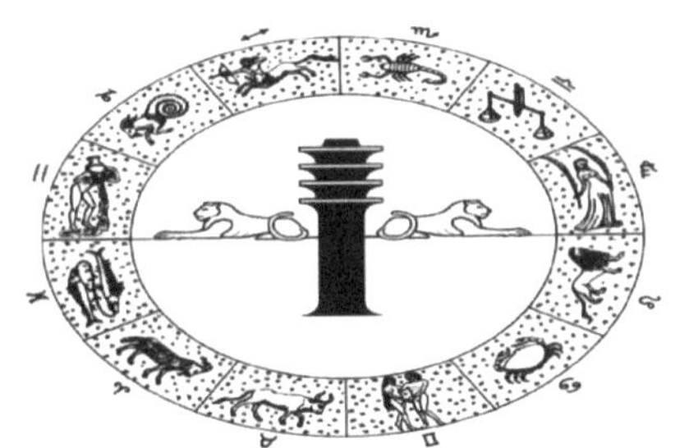

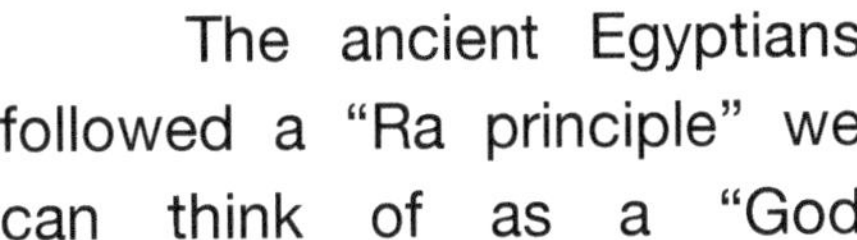

The ancient Egyptians followed a "Ra principle" we can think of as a "God particle". They understood aspects inherent in light that quantum physicists have only begun to understand in the last century. Although the manner in which they conveyed their stories in hieroglyphic form is quite different from the written language, the story they tell is one that we can understand. They understood that the creative force of light guided by the Ra principle creates the physical reality we perceive, and the absorption of more light creates the sentient experience of consciousness ascension.

The Ra principle is the God particle held within photonic consciousness, and within this physical world we can perceive ourselves and all matter as "frozen light"; meaning, light in its lowest vibration formed into matter.

CHAPTER THREE

PHOTONIC CONSCIOUSNESS: THE GOD PARTICLE

"Atum-Ra", Egyptian Son of God = Atom of God
"Adam & Eve", Genesis = Atom & Evolution, Gene of Isis
Isis being the female aspect of physical creation

All matter is comprised of vibrating atoms each unique in its creation. The atom is the smallest unit of a chemical element that retains its properties, and the molecule is the smallest collection of atoms of a given substance.[12]

Atoms are composed of protons having a positive charge and neutrons having a neutral charge that create the weight within an atom. Electrons carry a negative charge and orbit around the cells nucleus. Each of these atomic units carry information and are constantly moving and being energized by more quantum information. It is this miraculous movement that creates electricity and their moving currents create magnetism, which is observable as the electromagnetic field (EMF).

This trinity of electrical and magnetic forces, intricately dances within and around the atom to create all matter in this dense physical world, which we can think of as "frozen light'.

The photon (light) is the quantum information holder and carries information within its varying wave-like resonance in the multi-dimensional framework of the quantum information field.

---

[12] Rothman, Tony, Ph.D. 1995. *Instant Physics: from Aristotle to Einstein, and Beyond*: 67

Photons generate electric and magnetic fields that are the basis of the electromagnetism of our universe and all matter, including you and I. Photons have varying rates of vibration within the spectrum of light and move in waves as a color-coded harmonic frequency. The highest 'known' spectrum of light being gamma ray (violet in color), to the lowest being radio waves (red in color).

As the photon descends in the spectrum of light to its lower vibration (red) the energy is decreased and the wavelength is slower (longer) in this denser vibratory state. As the photon ascends energy increases and the wavelength is faster (shorter) as more quantum information exists within the higher spectrum of light.

The higher the spectrum of light and rate of vibration, the vaster and more expansive the field of information. It is as if the quantum information field expands to accommodate more light – photonic consciousness. Below is a basic illustration of the spectrum of light.

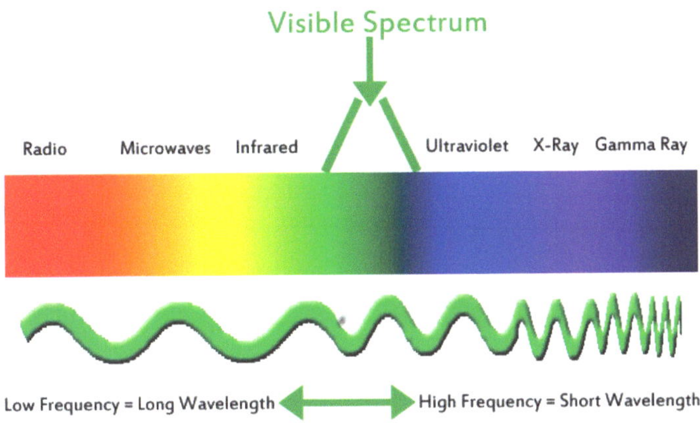

The photon is a conductor transmitting consciousness into the molecule & atom, and is the conductor by which consciousness ascends or descends from the force, or lack thereof, of its resonance (vibration). Photons are carriers of quantum information in what I have termed "Photonic Consciousness" in the multi-dimensional planes of existence. Within the physical world of third density, the conscious experience can be understood as embodied by three aspects; the physical, mental, and emotion, or "e-motion" meaning energy in motion manifest as feeling. Most religions refer to these three aspects as a trinity of mind, body and spirit, as Catholics would say, "Father, Son and Holy Ghost".

Consciousness ascension at the atomic level

Photonic consciousness residing in the quantum information field releases light energy, an exotic subtle energy, in harmonic, color-coded wavelengths of varying rates of vibration. As the photons descend the wavelength slows in vibration and less energy (information) is present. Conversely, as the photon ascends the wavelength vibrates faster as more energy (information) is present.

If a photon within a specific resonance field (vibration) collides with an electron, the electron can absorb the photon (light) and graduate to a higher quantum level. In this case the electron absorbs more cosmic energy-information from within the quantum information field of light, and thus ascends or graduates into a higher frequency. This ascension at the atomic level enables a higher influx of quantum information and with the continuous absorption of light the atom ascends higher and higher in frequency.

Science has proven that for the electron to ascend its resonance must match the energy of the higher quantum level, and the light wavelength as illustrated in the spectrum of light, will correspond to the color-coded frequency indicating its resonance.

Photons are the carriers of light and are moving points of energy that are surrounded by an electromagnetic field that give light its shape. The combination of many shapes in motion creates an electromagnetic wave. Not all shapes are of the same size. The difference in size determines the color of the light, whether it is within a high frequency or low frequency, and the wave determines the frequency by the size of the field. The high frequency fields, which contain more quantum information are more compacted in size producing a shorter wave, although these fields contain more quantum information. Conversely the lower energy fields, which contain less quantum information, are larger in size and thus the wavelength is longer and slower.

When light is propagated over long distances a gradual energy loss is incurred. The gradual energy loss results in larger EMF's and thus longer wavelengths. Therefore, could the "Central Sun" in the center of the Milky Way galaxy actually be higher frequency energy as a central source distributing quantum information through the galaxy?

All the light fields from gamma-ray (violet) descend towards the red lower energy frequency spectrum as a result of gradual energy loss. This results in a gradual stepping down in energy-vibration and slower wavelength. In this scientific concept I am reminded of the Biblical and religious stories of "Fallen Angels", with fallen meaning "stepping down" into the lower density of Earth.

As our Solar System transits through the plane of the Milky Way Galaxy and is immersed within the high frequency gamma ray bubble residing at its center, all atoms are ascending due to higher frequency energy penetrating its field, through the introduction of more quantum information. Our electromagnetic energy fields are becoming energized by this process and more quantum information within creation is causing our body, mind and Soul to awaken. In essence we are all becoming the

"Christ", meaning anointed with light. This will begin to make more sense as you view my photographic evidence and read subsequent chapters.

The Christ Consciousness

Quantum physicists have proven that as light energy is added to the electron it promotes the biomolecule - molecule – atom to the next quantum level. As light energy is decreased, the reverse happens. This light energy in this third dimensional universe is known as "biophoton", meaning photon of non-thermal origin in the ultraviolet spectrum of light emitted by a biological system (e.g. biological light).

Consider this, we see light within a very narrow band of the spectrum of light, however if we could see the entire spectrum of light in all its wavelengths simultaneously we would see the color "white". White is a pure form of quantum information comprising unity of all wavelengths, or unity of cosmic information. Thus higher dimensional beings we call Angels or Archangels, are often illustrated in ancient drawings or artwork in pure white sometimes wearing rainbow colored garments, reflecting their unity consciousness field. I call this the Christ Consciousness, meaning consciousness ascended in the light of God. Christ or Christos literally translated is "devoted" or "anointed". Could this be an ancient Greek reference meaning "anointed with light"?

Thus, the colors emitted by the quantum information field reflect the path of photonic consciousness within the field of the infinite Christ and Cosmic Consciousness.

On page twenty-three is a photo I took in June 2013 at the Church of Miriam in Axum, Ethiopia. This modern church was built next to the ancient church were it is believed the Ark of the Covenant once resided. In this photo you will see a sphere of white light often called "orb" which appeared at the feet of the painted image of the Virgin Mary and baby Jesus. Look closely and you will see a rainbow of color that radiates off the left and right of the sphere. Look even closer and you will see what appear to be wings and delicate white, etheric arms extending from the sphere reaching up towards the image of the Archangel Gabriel (messenger Angel). As I will explain, this picture captures the "biophotonic energy" of photonic consciousness. Notice the painting depicting rainbow colors and the Archangels Michael and Gabriel painted to the left and right holding the metaphorical "light sword of truth."

On the right is a cropped and zoomed perspective for closer viewing.

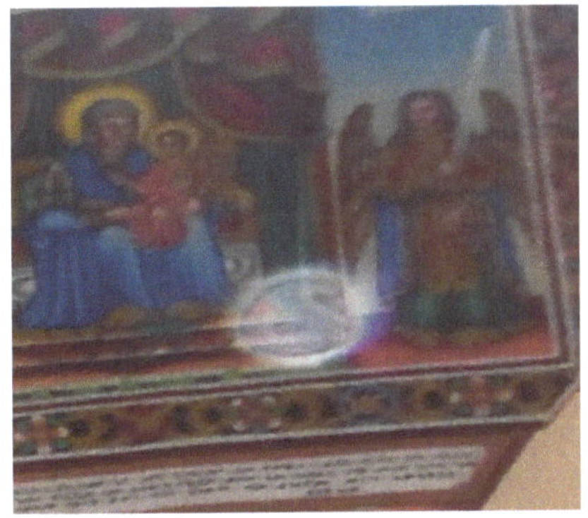

What is black?

With this understanding of light and color, what is black? Black is the antithesis of white. Black resides at a very low frequency and wavelength where no light exists. If there is no light you will see or perceive "black". Black in its low frequency, lacking light, has very little quantum information coming from the creative force of God. Thus we can understand that in the low frequency of black wherein little quantum information resides exists the metaphorical meaning of the "antichrist", or the "unconscious". The antichrist is that which is lacking in light and lacking in the Christ Consciousness. This is a space within the quantum information field residing outside of the knowing of God. However, this darkness can dissipate instantly with the introduction of light.

Along the spectrum of light there are endless levels of resonance, and a simple introduction of "light" creates the immediate dissipation of dark as the wavelength/information is introduced. As we know, science has proven that molecules and atoms do absorb specific frequencies of light, and white light comprises all wavelengths. As I propose, this is the pure form of cosmic unity consciousness – the Christ Consciousness.

Ascension of the Atom

In the concept of ascension at the atomic level, electrons have energies that are quantized in specific energy levels - resonance vibrations. The electrons can undergo transitions from one level to another, and as I have explained, ascend by the absorption of more light. Where science fails to understand the nature of the conscious universe, the quantum information field of God, is by avoiding the discussion of what I have termed, "Photonic Consciousness", represented simply as "C". It is this subtle energy in the unseen world that quantum physics is just beginning to explore; a unity field of information that animates our physical world and affects the laws that govern it.

I believe this unseen world residing in a multi-dimensional framework outside of the space-time continuum, is the world from which our Soul came and is the world to which are Soul returns.

On page twenty-five is a picture I took of photonic consciousness appearing as a "biophoton" that materialized at my bedframe in August 2013 during a period in my life when I was trying to understand the nature of the divine experience I was having. I had the intuitive knowing that the spheres of light I was capturing in photograph were illustrating the story of creation, but until the sphere with the "C" appeared in my photograph did I begin to connect-the-dots. This picture below represents a profound moment in my awareness of the creative expansiveness of light. It is from this moment that the cosmic wheel and journey of Photonic Consciousness was a story I knew I had to share.

In the photograph below, to the left is the original taken near my bedframe and on the right is a cropped/zoomed perspective for closer examination. You will notice in the photo there is the letter "C" inside the sphere. Interesting coincidence, a crop circle appeared in Johnson City, Tennessee May 13, 2013 forming the same "C" shape with a dot in the center of the "C". I was unaware of this crop circle until it was brought to my attention by a friend.[13]

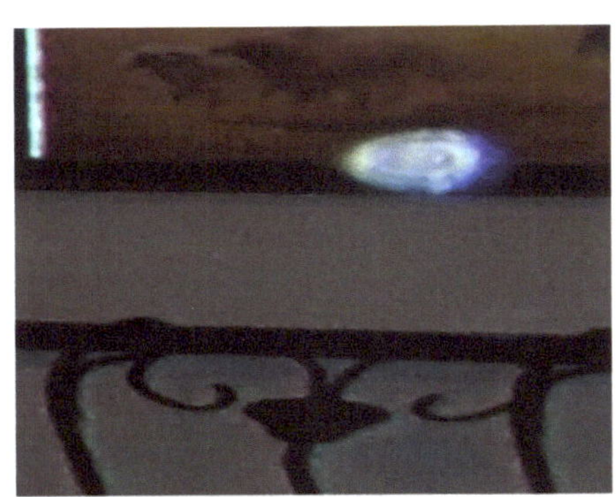

[13] www.cropcircleconnector.com

CHAPTER FOUR

SUBTLE ENERGY

"Gravitation is not responsible for people falling in love."
Albert Einstein[14]

MIT researchers are close to proving that the laws governing our universe are not solely based on Newtonian physics (also known as Newtonian mechanics). Isaac Newton (1623-1727), whose theories became known as Newtonian physics, expanded on Galileo's work to define the relationship between energy and motion using the dynamic concepts of acceleration, momentum and conservation laws. Isaac Newton won widespread acceptance of the modern concept of a scientific law with his three laws of motion and his law of gravity, which accounted for the orbits of the earth, moon, and planets, and explained other phenomenon such as the tides.[15] In modern times, most scientists will tell you that the Law of Nature is a rule based upon an observed regularity, and these laws are defined in mathematics. Most scientists will tell you that if you cannot observe it, then it is not provable in scientific terms.

However, quantum physicists are stretching this boundary that has governed science for centuries into the realm of the

---

14 http://www.brainyquote.com/quotes/authors/a/albert_einstein.html#iyqOmSo8cDdU2PHl.99

15 Hawking, Stephen & Mlodinow, Leonard. 2010. *The Grand Design.*

unseen, subtle quantum energy field. This quantum view may seem improbable and even counterintuitive to accepted laws of the universe, however it does illustrate the miraculous complexity of such phenomena as entanglement. Entanglement is a process by which one particle instantly affects another faster than the speed of light despite these particles being at opposite ends of the universe. It is within the field of quantum mechanics that scientists are now realizing that the universe is far more complex than previously understood, and there exists a quantum force of unseen energy, a quantum information field, that has a dramatic impact on everything. MIT researchers have concluded there are "hidden variables" that give rise to an unseen quantum field.

> As MIT researchers elucidate, *"quantum physicists refer to "setting independence," or more provocatively, "free will." This loophole proposes that a particle detector's settings may "conspire" with events in the shared causal past of the detectors themselves to determine which properties of the particle to measure — a scenario that, however far-fetched, implies that a physicist running the experiment does not have complete free will in choosing each detector's setting. Such a scenario would result in biased measurements, suggesting that two particles are correlated more than they actually are, and giving more weight to quantum mechanics than classical physics."*[16]

Or, could it be that the particles being measured responding to the conscious intention (thought process) of the physicist react in accordance to a quantum information field of

---

[16] MIT News: "Closing the Free Will loophole: MIT researchers propose using distant Quasars to test Bells theorem." February 20, 2014

its own, not necessarily 'conspiring' against the free will of the physicist but rather consciously choosing to express itself in accordance to its own free will. Could it be that what the physicists have discovered is proof that a conscious guiding force exists within particles?

This is indicative of an experiment done by a scientist measuring the consciousness of plants. From my first book *Smoke and Mirrors* I use the example of a scientist who used a polygraph machine and attached it to a plant to determine if a level of consciousness existed, which could be measured by its electromagnetic energy field. To test the theory the scientist did harm to the plant, such as tearing a leaf, but the plant's energy field didn't react. The scientist then tried other physical methods to harm the plant, but again nothing. Then the scientist sat at his desk and simply thought deeply about using a lighter to burn a part of the plant. The result was instantaneous. The graph on the polygraph machine began jumping in a manner akin to a human caught lying. The plant was indicating that it was under stress. The scientist tested this theory again and soon learned that the thought of a negative action (a low frequency vibration) directed at the plant invoked a low vibratory (fear-based) response from the plant. It was as if the plant received the low frequency vibratory thought, an energetic resonance, and responded to it in-kind. At that point when the scientist physically approached the plant with the lighter, the plant continued to panic. It was if the plant continued to receive the low frequency energy and responded according to past experience.

Another experiment done to illustrate the power of consciousness imbued within our thought and the frequency of its resonance is from Masuro Emoto's book *The Secret Life of Water*. In this book the author conducts a series of experiments on water using conscious intention, e.g. directed thought. What the author found after numerous controlled experiments is that indeed the water molecule responded in direct correlation to conscious intention. In cases where the intention was positive, of a high frequency, the water molecules frozen as snowflakes took on the form of beautiful, perfect geometric formations reflective of the intricacy of higher frequency harmonic vibration (pitch). Where the intention was negative, of a low frequency, the snowflake had broken and distorted geometric shape. In both cases, consciousness had a direct impact on matter.

The Unseen Creative Energy

Quantum physicists are exploring an unseen creative energy often referred to as "subtle energy". It has been theorized that subtle energy comprises the quantum information field, an unseen force that does not function within the constraints of Newtonian physics. Subtle energy may be the network of information by which photonic consciousness travels within the multi-dimensional universe(s). Physicists are beginning to explore the idea that this unseen force of subtle energy may lead to proving the existence of a God particle.

Subtle energy does have a dramatic impact on measurable forces such as electromagnetic energy, nuclear energy, gravity, mass and atomic bonds. Although modern quantum physicists perceive to be on the cusp of "discovering" this energy within the matrix of creation, it has been known and understood for thousands of years. Subtle energy has been known and venerated as the creative life force by many names; prana or breath, love, kundalini, aura field energy, consciousness, ether/aether/eter, energy of intention, merkaba, chi and ki.

Through my journey of exploration and discovery to better understand this extraordinary truth, I define subtle energy as the matrix of the quantum information field wherein infinite photonic consciousness transits in a multi-dimensional framework of creation bringing forth the creative force governing and animating all things in all dimensions. The Prime Creator, God, is the source of photonic consciousness, and you and I are aspects (or children) of its conscious creation. We are lesser co-creators with the Creator in this grand cosmic and infinite tapestry of light and sound.

Subtle Energy

To understand what subtle energy is let us look at other aspects of energy. There are four aspects of energy existing within the third dimension I will call simply "e". Potential energy, which keeps an object in its state or position, kinetic energy which keeps the object in motion, thermal energy which is heat in various forms, and mechanical energy which is the energy an object has as a result of its position or motion.[17] These various forms of energy, "e", are well known and understood by scientists and engineers. "e" can manifest in the positive space-time continuum, is electrical in nature and has a positive mass.

"e" travels slower than the speed of light and gives rise to the force we call gravity. Furthermore, "e" is based upon the scientific law whereby opposites attract and similar electrical charges repel. It is the endless dance of attraction of duality (opposites), and it is fitting that this dance take place in the physical third dimension as this is a place governed by light and dark, positive and negative.

Conversely, subtle energy I will refer to as "se" occupies a negative dimension and has negative mass, meaning no mass at all, thus it is truly unseen. It can reside outside of the space-time continuum, but yet it can exist within the space-time continuum. As a massless energy it moves freely in a multi-dimensional framework. Subtle energy is magnetic and travels faster than the speed of light and can hold unlimited amounts of quantum information. Therefore being magnetic in nature based upon

[17] Rothman, Tony, Ph.D. 1995. *Instant Physics: from Aristotle to Einstein, and Beyond*

vibration and harmonic wavelength resonance (sound), like attracts like. Subtle energy arises from the electromagnetic wave whereby light is vibrating at varying wavelengths and speeds. This movement and sound define its field of resonance within the spectrum of Photonic Consciousness and tells the observer the story of the Soul's journey with light being a demonstration of quantum information and experience(s). The photon/biophoton is the carrier of electromagnetic radiations at varying wavelengths, hence Photonic Consciousness.

"E" versus "SE"

"E" = energies	"SE" = subtle energy
Exists in third dimension	Is multi-dimensional
Electrical in nature	Is electromagnetic in nature
Has a positive mass	Has a negative mass
Travels slower than the speed of light	Travels faster than the speed of light
Gives rise to the force of gravity	Can hold unlimited amounts of quantum information
Opposites attract and likes repel	Like attracts like

> *"Stanford University professor [of materials science and engineering] Dr. William Tiller is an expert on subtle energy. Dr. Tiller states subtle energies are not measurable by physical means, but we can detect their signals. This is because as subtle energy changes from one form to another it creates a transducer signal at the magnetic vector. Subtle energy also generates electricity and magnetic signals that have observable effects. Tiller states the following about subtle energies:*
>
> - *Subtle energy can be manifested by*

people and directed by the flow of energy through conscious intention.

- *The mind-electron interaction is effective even over great distances and the time/space continuum.*
- *Subtle energy occupies a different time-space domain. These domains are different levels of reality that flow down to the physical (dense) world from the highest domain Dr. Tiller calls the "Divine". Each domain provides the template for the domain below, and adapts and instructs with information. The laws in each domain differ based upon density. [Thus, the laws of the physical domain may be Newtonian in nature, but this is not the case for the laws of higher domains within subtle energy.]*

The density levels, as defined by Dr. Tiller, from least dense (being 1) to most dense (being 4).

1. *Divine – the quantum information field, which is fifth dimensional and above.*
2. *Astral – fourth dimensional*
3. *Etheric - bioplasmic, pre-physical or energy body [biophotonic]*
4. *Physical - third density*

For subtle energy, as time passes in this domain, potential increases and entropy decreases which allows more order to be established and less chaos. In the physical world it is the opposite. As time passes in the physical domain entropy increases thus giving rise to "chaos". Dr. Tiller states that the subtle energies can be observed through intuition – the sixth sense [i.e. 6th chakra – Third Eye]. Tiller states that the meridians and

> *chakras operate like antennae that detect and send signals from the physical into the upper domains of subtle energy. These subtle structures interact between the physical body and etheric realms and above, illuminating higher orders so that we can perceive them from the physical plane. Dr. Tiller proposes that human consciousness occupying the same vibrational space as subtle energies will attract this energy into the physical plane and shift reality to the information contained within that wavelength.*[18]

As quantum physicists attempt to unlock the secrets of the cosmos within the realm of the quantum information field of photonic consciousness, they postulate and use means and methods based upon the paradigm of Newtonian physics. As Dr. Tiller proposes, this method of exploration will not produce the results they seek, but rather subtle energy must be explored from the realm of consciousness and conscious intention. This is in fact the personal experience I have had.

It is only within our co-creative aspect of photonic consciousness that the gateways to knowledge and understanding of this ultimate creative force coming from God will be felt and observed.

Evidence of Photonic Consciousness in the Subtle Energy Field

Several years ago I was propelled on a journey of discovery and exploration into the spiritual realm of unseen subtle energy. For reasons I have personally surmised, I have been blessed to capture over 4,000 photographs and

---

[18] Dale, Cindi. 2009. *Subtle Body: An Encyclopedia of your Energetic Anatomy*. 2009: 12-13

documented evidence of photonic consciousness and its creative force that lifts the veil between this world and higher dimensional planes of existence residing outside time and gravity. Is this divine intervention or conscious intention? Is there any difference between the two? I will let you decide.

Throughout the pages of this book I will present a small selection of pictures I have taken since 2012 in my home and places around the world. For purposes of methodology, I have captured these photos with different cameras, taken pictures during daylight and nighttime under different lighting conditions, taken photos capturing this evidence in the presence of others, taken the photos with and without the camera flash, and in some cases random strangers took the photos for me with my various cameras. I have all photos on camera SIM cards and challenge any forensic photography expert to examine them for authenticity, and in no case have I done any editing other than crop and zoom for closer analysis of the sphere(s) and in some cases overlay on the photo a blue parenthetical to point out the sphere, as many spheres are present in some photos.

In all cases the photonic consciousness is unmistakable and clearly visible. As you will see in the photographs, the spheres, some call “plasma orbs”, appear in varying geometric formations from a circle, sphere, oval, rectangle, triangle, hexagon, octagon, cylinder, cube, nonagon, decagon and so forth. As a further illustration of the nature of complexity, the photonic consciousness appears in many colors often radiating the spectrum of light from opposites ends, and contains within its center evidence of human-like faces, letters, wormhole(s) and wave patterns imparting powerful messages and meanings to the observer.

I will begin with photonic consciousness, which I will refer to interchangeably as "sphere", that appeared during the afternoon on a book I read the summer of 2013, which helped to ignite my understanding of the multidimensional experience of the Soul, called "*The Alchemy of Nine Dimensions*" by Barbara Hand Clow. In the photos to the right, notice how the sphere moved from one book to another on the same bookshelf. The sphere moved to a scholarly interpretation of ancient Egyptian hieroglyphics detailing the mythology and teachings of Thoth. As I will show you, the spheres often communicate to the observer in extraordinary ways.

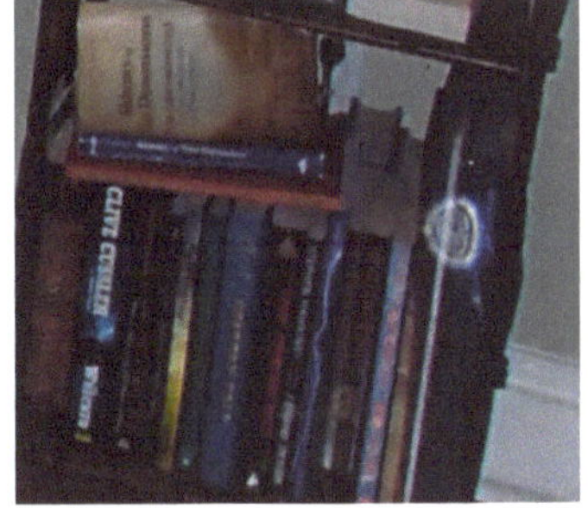

I took the photo on page thirty-six one evening in my family room. I noticed my parrot, Coco, observing and responding to an unseen energy in the room. I could intuit the presence of this energy, but could not see it with the naked eye. I picked up my camera and took the below photograph on the left. On the right is a cropped/zoomed perspective for closer examination of the shape and light radiating from the photonic consciousness. As you will notice on the top right, it is shaped like a circle/sphere. On the below right is a cube shape that illustrates the varying ends of the spectrum of light resembling a 'doorway' or gateway.

I have pasted a sun over my husband's face for privacy and pasted parentheticals to identify the Sphere and Cube cropped/zoomed to the right.

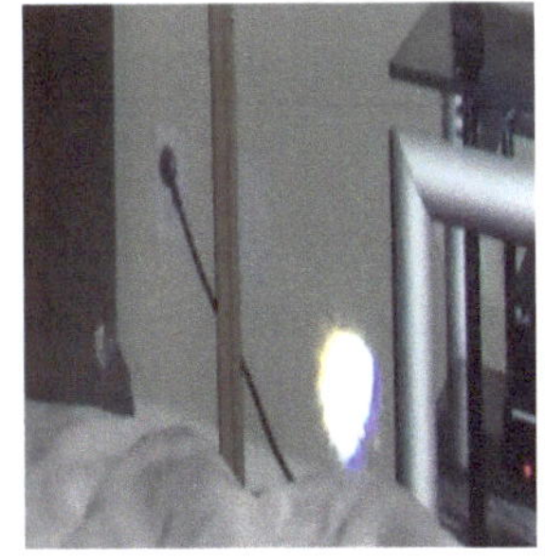

On the right is a photo that was taken in my bedroom where the upside down cone-shaped sphere appeared near the wall, again illustrating the spectrum of light as a stargate. You can see etheric subtle energy being released into the room from the sphere, which appears to be a doorway into another dimension.

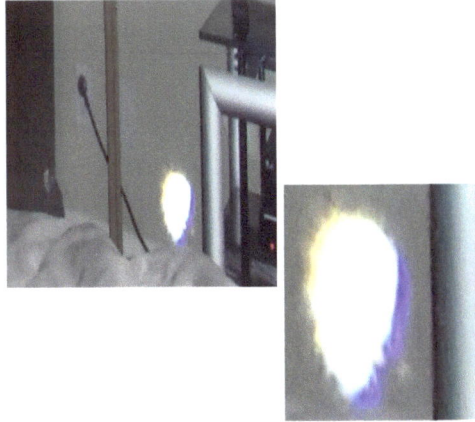

The next photograph, below, was taken of me meditating inside the Temple of Osiris, Room of Ra, in Abydos, Egypt March 2013. Notice the dance of light and stargate/wormhole opening at the top of the photo.

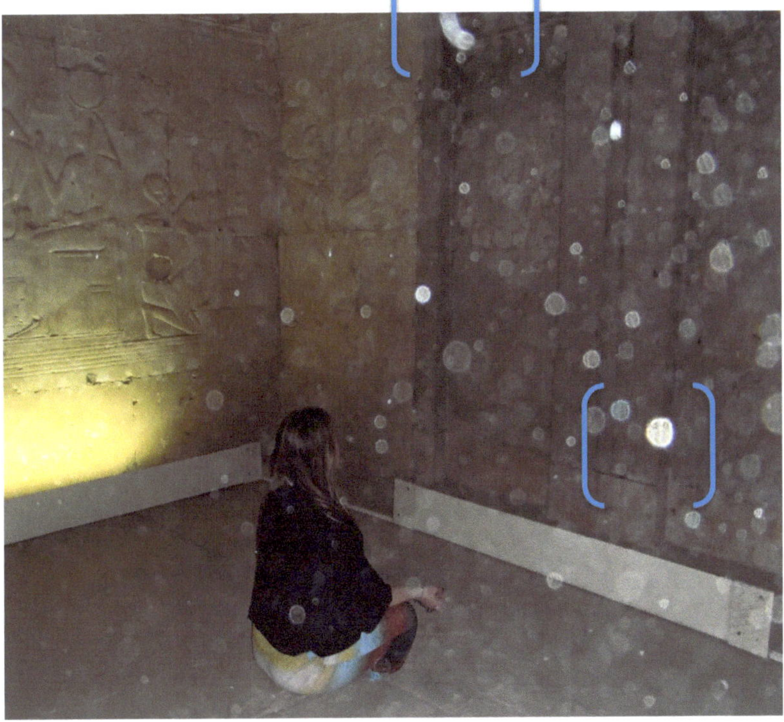

Below is a cropped/zoomed perspective of the wormhole and a sphere for closer examination. In the cropped photo on the left the wormhole takes the shape of a cylinder appearing as a stargate/wormhole opening, and you can even see subtle energy being released into the room. On the right inside the sphere you will see a cross and 'beings of light' on opposite sides of the cross.

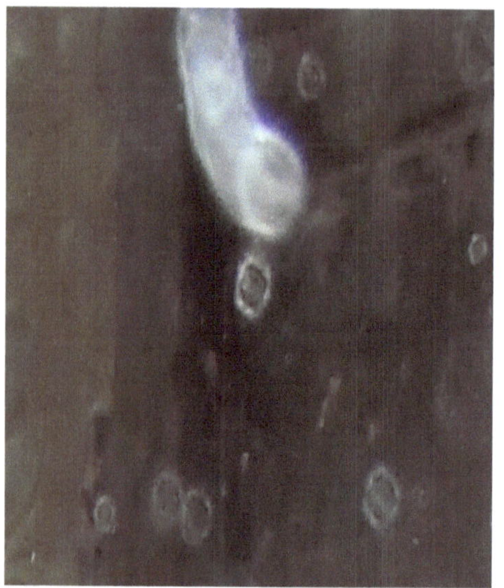

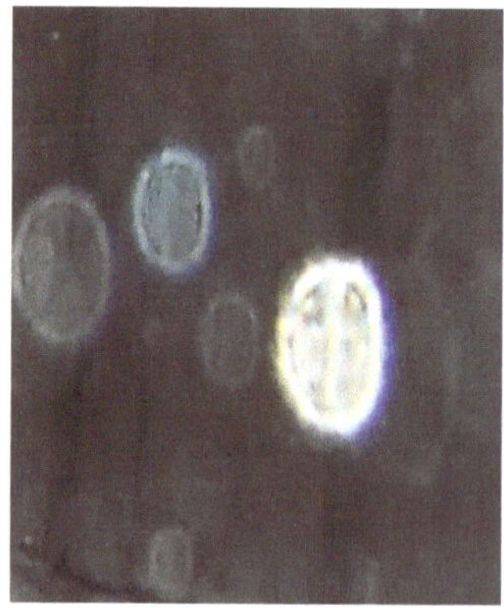

I took the photo on page thirty-nine inside the Room of Thoth in the Temple of Osiris in Abydos, Egypt same day. The photo on left is the original, and the photo on right is cropped/zoomed to provide closer examination of the expanding stargate/wormhole that resembles the shape of a pyramid, and appears to be releasing subtle energy into the room.

I took the photo, below, in my bedroom in 2014. Notice the violet spheres in a geometrical formation illustrating the Vesica Pisces, overlapping spheres, which I will discuss in detail in this book.

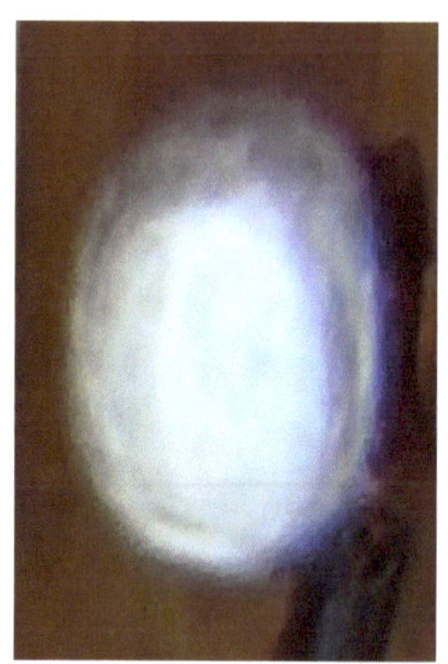

The above photo was also taken in my bedroom of a sphere that appeared on a painting of a sphinx hanging next to my bed. In the cropped photo, to the right, you can see a face inside the sphere.

On pages forty-one to forty-six are a series of photos I took during my June 2013 trip to Jerusalem. During that trip I captured incredible photonic consciousness at the feet, head and hands of Jesus, the Virgin Mary, Mary Magdalene and images of the Archangels Michael and Gabriel. The first picture I will present was taken on Mount Olives at the Church of Gethsemane in Jerusalem, and is a painting at the church alter of the "rock of agony". This is believed to be the location where Jesus stood the night he was arrested by the Romans before his crucifixion. The significance of this location, it is the area where Jesus stood on a rock in his final hours on Earth before he transfigured into "light". From the Gospel of Mark we learn of 8 incidents occurring from the Last Supper, each being 3 hour intervals (a trinity) to Jesus' crucifixion. From the time of Jesus' arrest to his crucifixion was 12 hours. In numerology the 12 is 1+2 = 3, again the trinity. As you will notice in my photos often the spheres position themselves in a "V" formation of three, a trinity.

As you will see in the photo below and on the next page, the spheres are at the feet of the image of the Archangel Gabriel and the feet of the image of Jesus who is sitting on the "rock of agony", a literal rock that is just below this painting at the church alter. The spheres continue from this painting to form a "V" shape connecting to another painting on an adjacent wall depicting Jesus speaking to his Apostles. This is the exact location where in early 2008 I was standing when a powerful energetic force I can only describe as a "wind" swept over me and brought me to my knees in emotional pain, an emotional agony. I began crying uncontrollably and in that moment began to understand and awaken to the power of the Divine and magnificence of God. It was the beginning of a profound awakening of my consciousness, and as I embraced my awakening the information contained within the photonic consciousness accelerated exponentially.

The photos below are cropped/zoomed perspectives of the original photo on page forty-two. The spheres depict a powerful message of the Divinity of Jesus and the story of his last days on Earth.

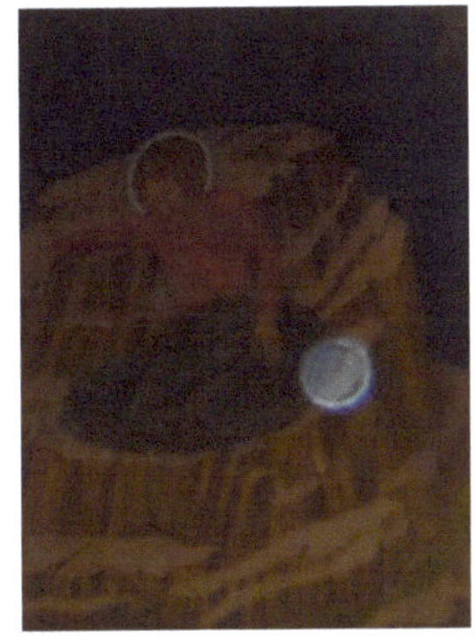

As I previously mentioned, the above photo is the original which I took in the Church of Gethsemane in Jerusalem. You will notice in the painting on the wall, the artist painted white halos of light circling Jesus and the Archangels Gabriel's heads. The similarity of the white light halos in the painting, to the spheres that appeared in the photo as the feet of Jesus and the Archangel Gabriel, is incredible. Clearly the spheres are illuminating a powerful message and directing the observer to acknowledge Jesus' divinity.

The photograph on page forty-three was taken inside the Church of Mary Magdalene in Jerusalem on the same trip, not far from the Church of Gethsemane. I had the opportunity to attend mass in the church and take photographs afterward. In an amazing display of light, as seen in the photograph, the spheres appear near to Jesus, and at his hand as he blesses Mary Magdalene. To the left is a zoomed/cropped perspective of the original photo to the right.

In Jerusalem is the Church of the Holy Sepulchre, also known as the Church of the Resurrection. It is a magnificent church and site of pilgrimage for Christians who wish to pray at the tomb of Jesus, the Christ.

While in Jerusalem I visited this beautiful Church and was blessed to capture an extraordinary photograph of Divine illumination on Jesus' tomb. On page forty-four you will see what for me is one of the most magnificent photos I have in my collection. In the photograph you will see the illumination of the spheres beginning on Jesus' tomb (colors of green and blue), followed by the ascension of the spheres in a kaleidoscope of color to the dome of the Church. The "dome" of the Church architecturally represents the upper half of a sphere and cosmic divinity. In this context, the word "dome" in Greek is "Domus Dei", meaning "House of God".

I took the below photo in Jericho on the same trip June 2013. This artistic sculpture was in a monastery. Notice the location of the spheres almost like 'thought bubbles' from the image of Mary Magdalene to Jesus.

I took the below photo in the Church of the Virgin Mary in Jerusalem June 2013.

I captured a magnificent photo in August 2013 close to the same period when the “C” sphere appeared at my bedframe. I had just finished meditating and could feel a divine presence in the room. It is a feeling I have become accustomed to recognizing. I noticed my dog observing unseen energy on the bed, and I pointed my camera and took an incredible photograph. In the photo, on page forty-seven, is an extraordinary message conveyed by photonic consciousness in the form of a white, etheric horse. You will see in the photo a clear and visible image of the face, mane and neck of a white horse inside the sphere of light. This horse has significant meaning for me in my personal spiritual journey.

It was in 2011 I began having vivid, lucid dreams of a white horse coming to me. This happened night after night, and the white horse in my dreams was very persistent in getting my attention. When I woke up, I had a clear memory of my dream, and the white horse, which I lovingly named "Pegasus".

In January 2012 I wrote my first book *Smoke and Mirrors*, and in October 2012 I wrote my second book *Approaching Singularity: the Genesis of Creation*. After having written these books the acceleration of my consciousness increased dramatically.

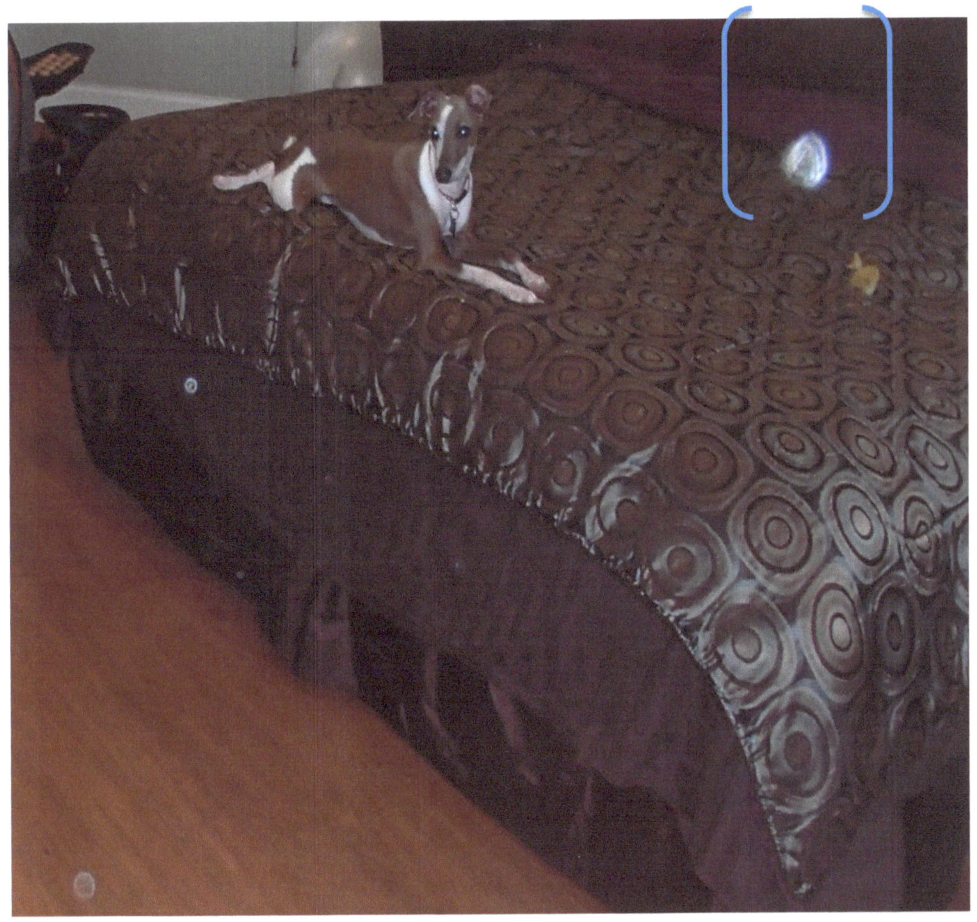

As I journeyed along the path of my spiritual awakening, I became a metaphorical "empty cup", and allowed the Christ Consciousness to fill my body, mind and Soul with Divine wisdom and truth. As I accelerated along this path, I began to have an understanding of subjects I have no academic degree in, nor had I previously explored. For me now, in 2014, the appearance of the biophotonic, subtle energy of photonic consciousness is an everyday occurrence and their message(s) has become more complex and profound.

The below picture was taken of me inside the King's Ascending Chamber in the Great Pyramid of Giza, Egypt, December 2012. I had traveled to Giza to experience the incredible alignment of Mercury, Venus and Saturn on December 3, 2012 with the three pyramids of Giza. This is an event that is believed to occur only once every 2,737 years. The original picture is on the left taken by a pyramid guard, and cropped/zoomed perspective of the sphere on the right. Do you see the Being of light inside the sphere? To me, the Being of Light resembles a butterfly with wings spread.

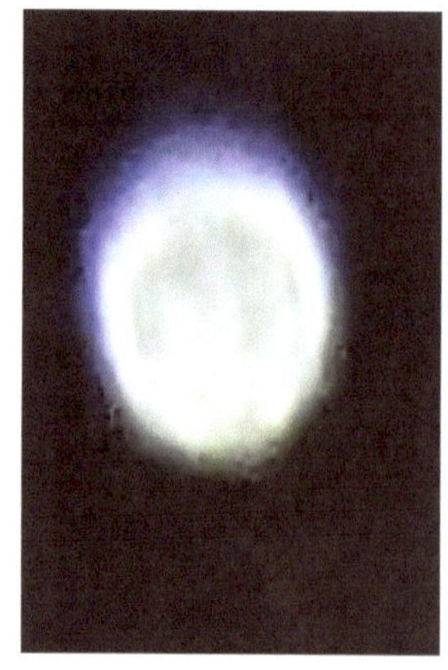

In gazing at these photos I recall a passage I once read from a book published in 1934. At the time I did not understand the meaning, as I do now. I want to share the passage with you as it eloquently expresses the power of consciousness. From the experiences of my journey I have come to understand that we are co-creators with the Prime Creator, God, and as quantum, infinite consciousness we are merely having a human experience. We are the mirror image or "picture" of the divine consciousness that created us and through free will and free choice can choose to participate in the co-creative experience of our reality rather than allow ourselves to be a victim of its circumstance. This is a divine gift bestowed to us by God, similar to the gift we bestow upon our children as we bring them into this world, and nurture them from childhood into adulthood.

In this moment along the path of the cosmic wheel, humanity is the cosmic butterfly awakening from our cocoon rediscovering our place in the multidimensional universe.

> *"Always remember you are God picturing. You are God Intelligence directing. You are [a] God Power propelling. It is God's, Your Substance, being acted upon. As you realize this and contemplate the fullness of it often—everything in the Universe rushes to fulfill your desire, your command, your picture, for it is all constructive and therefore, agrees with the Original Divine Plan for Self-Conscious Life. If the human side of us really agrees to the Divine Plan and accepts it—there can be no such thing as delay or failure, for all energy has the inherent quality of, Perfection, within it and rushes to serve its Creator."..."As your desire or picture is constructive, you are God seeing His [Her] Own Plan.*

When God Sees, it is an irrevocable decree or command, to appear Now. In the creation of this earth and system of worlds, God said, 'Let there be Light', and Light appeared. It did not take aeons of time to create Light. The same Mighty God is in you, now, and when you see or speak, it is His [Her] attribute of sight and speech, which is acting in and through you."..."If you realize what this truly means, you can command by His [Her] Full Power and Authority, for you are His [Her] Life Consciousness, and it is only the Self-Consciousness of your Life, that can command, picture, or desire a constructive and Perfect Plan. Every constructive plan is His [Her] Plan. Therefore, you know God is acting, commanding, 'Let this desire or plan be fulfilled now', and It is done. What one thinks upon, he [she] draws unto himself [herself]."[19]

[19] King, Godfre Ray. 1934. *Unveiled Mysteries*: 110

CHAPTER FIVE

VESICA PISCES

"As Above, so below, as below, so above"
Hermetic Axiom

The "circle" is a powerful symbol in mysticism, religion, ancient culture and tradition, science and mathematics. It is symbolic of the infinity of creation, the alpha and omega under the guiding force of God. It represents the idea of continual existence and the perfection of creation.

When you overlap a second circle you expand the meaning of perfection into one of duality with aspects of positive/negative, light/dark, male/female, god/goddess. The overlapping spheres become symbolic of the god/goddess aspect of Prime Creator, God, which brings forth physical manifestation and creation. This mystical symbol known as the "Vesica Pisces" can be seen throughout religious depictions of creation, rendered magnificently in artistic work, and is a Rosetta stone memorialized in history and culture pointing humanity towards the relationship between infinite consciousness and creation.

As mentioned, as a second circle is added it expands this unity to give rise to duality, meaning light and dark, positive and negative, male and female; and, it is by overlapping the two spheres that a unity emerges bringing forth creation. The concept of the Vesica Pisces is mathematical and a fundamental principal of geometry. It is the essence of the symbol's of the Flower of Life and the Tree of Life with its meaning of creation dating back thousands of years to ancient Egypt and India. The geometry of the Vesica Pisces illustrates the cycle of world ages and cosmic wheel of time. This is the foundation of the astrological cycles, and the precession of the equinoxes wherein the Vesica Pisces represents the summer and winter solstices within the cycles of time.

Electromagnetic Field and Sacred Geometry

The image of the Vesica Pisces is evident all over the world, hidden in plain sight on doors, windows of churches and temples, and ancient monuments. The cross and circle in the form of the Celtic Cross and Sun Wheel date back thousands of years, and these symbols together form the key to understanding the cosmic clock held within the cosmic wheel. It is this cosmic clock that ushers in dramatic change for our solar system, planet and all life residing on it through the introduction of more quantum light information (Photonic Consciousness) or lack thereof. In this concept we can understand the epochs of golden ages and the decline into dark ages.

In the space of the overlapping circles resides an electromagnetic field, which is located at its center as illustrated by the cross in the graphic to the right.

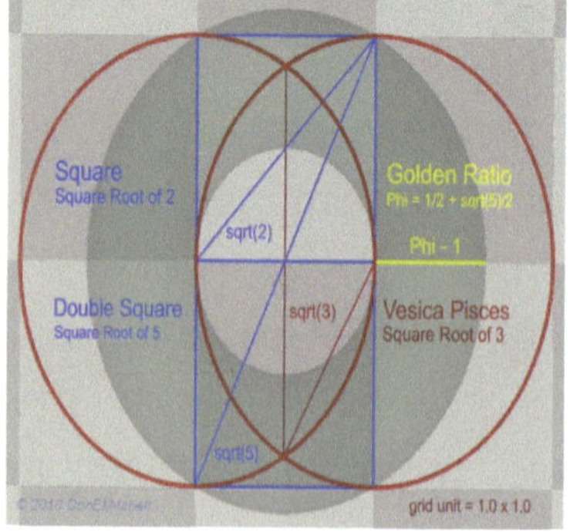

The Vesica Pisces and Golden Ratio of Phi's geometry is found in all living matter, which makes this symbol a Rosetta stone for humanity to discover the nature of creation.

Phi is the mathematical proportion (1.618) of the golden ratio (golden mean), also known as “divine proportion”. Just as “pi” is the ratio of the circumference of a circle to its diameter, phi is the ratio of the line segments that result when a line is divided.

It was circa 300BC when a Greek mathematician, Euclid, in his "Elements" divided a line at the 0.6180399.. point, dividing a line in the extreme and mean ratio. This later became known as the golden mean. This golden ratio can be seen in the Great Pyramid of Giza and other ancient architectural wonders, and depicted in some of the most famous works of art.

In the 16th century, Luca Pacioli wrote a book called “*De Divina Proportione”* (translated as The Divine Proportion). His book expounds on the drawings of Leonardo da Vinci and the 5 proportions. Just as the golden ratio of phi can be seen in all physical creation, so to is the Fibonacci sequence.

The Fibonacci sequence is a set of numbers that starts with a one or zero (followed by a one) and proceeds based on the rule that each number is equal to the sum of the preceding two numbers. For instance, if the Fibonacci sequence is denoted $F(n)$, where n is the first term in the sequence for 0, you have the following equation:

$F(\mathrm{n}) = 0, 1, 1, 2, 3, 5, 8, 13, 21, 34 \ldots$

The Fibonacci sequence is derived from successive numbers and dividing them by the next number in the sequence, and as you do this the ratios get closer and closer to the Golden Ratio Phi 1.618034. This mathematical sequence is inherent to the mathematical proportions of all living matter, including DNA, our solar system and the Milky Way galaxy. In this aspect math can indeed be understood as a language of God, and a set formula for physical, proportional creation.

Leonardo da Vinci's famous drawings and paintings illustrate this wisdom understood centuries ago. Notice the similarity of the spiral with the galactic spiral and in the center resides the Vesica Pisces.

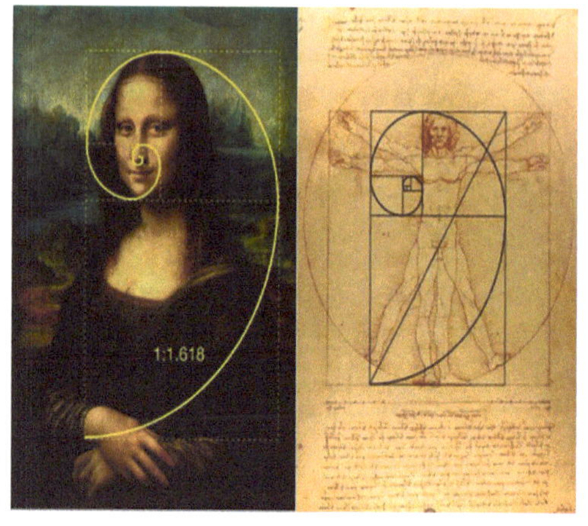

The Astrology of the Vesica Pisces and Cosmic Precessional

The "winter solstice" of the cycle of Age's took place in the epoch circa 10,000 BC during the astrological Age of Leo, the lion. We now approach the "summer solstice" point along the cosmic wheel in the astrological Age of Aquarius. This is the cosmic precessional governing the astrological world ages.[20]

This precession courses the celestial pole to trace out a cone round a central axis, of which the earth's axis is at a 23.5-degree angle to. As the celestial pole moves round this axis, the celestial equator moves relative to the ecliptic, the points of intersection of these two

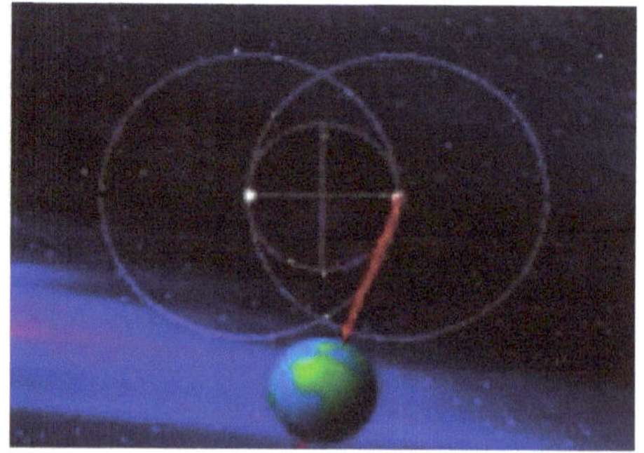

Notice the Vesica Pisces represents the electromagnetic center of concentrated energy corresponding to the photon belt and "central sun" located at the center of the Milky Way galaxy. "As above, so below, as below, so above."

[20] http://www.vesicapiscis.co.uk

circles, the equinoxes, move along the ecliptic from Age to Age, about 1 degree every 65 years. The slow change in the orientation of the earth's axis of rotation courses the sun to rise with a different constellation on the vernal equinox every 2,000 years. What this means is, on the vernal equinox of the yearly cycle of the zodiac in the Age of Aries, the sun rises with Aries as backdrop, in the Age of Pisces, with Pisces as a backdrop, in the Age Aquarius with Aquarius as a backdrop and so on.[21]

For the past 2,000 years the sun has been rising with Pisces as a backdrop, and now for the next 2,000 years as we transit through the photon belt on the plane of the galactic center our sun is rising with Aquarius, hence the Age of Aquarius.

The last 12,500 years of world ages has marked shifts in consciousness as our solar system transits from its 'winter solstice' (dark age) to its 'summer solstice' (golden age).

- Circa 10,000 BC Age of Leo (Lion)
- 4,000 BC Age of Taurus (Bull)
- 2,000 BC Age of Aries (Ram)
- 0 AD Age of Pisces (Fish)
- 2,000 AD Age of Aquarius (Water)

If we look to the culture of ancient civilizations during these epochs along the cosmic wheel, we find a remarkable story corresponding to the astrological clock. Very little is known about the age of Leo circa 10,000 BC. Some authors such as Graham Hancock writing about the hidden history of humanity in his popular book "*Fingerprints of the Gods*" details the historical and archeological evidence that the Great Pyramid of Giza and Sphinx (face of a human and body of a lion) was in fact built in the Age of Leo (lion) some 12,500 years ago and that ancient Egyptians inherited this archeological wonder. Hancock cites

[21] http://www.vesicapiscis.co.uk

geological evidence such as water erosion and other factors to make this extraordinary case.

In the epoch of 4,000 BC, the Age of Taurus, the civilization culturally centered around the symbol of the bull, with Apis Bull being the incarnation of Ptah of Memphis, or Hathor the goddess of Earth symbolized by a cow.[22] The Age of Taurus is the age of the female goddess and an age of the divine feminine – the Inanna (mother). In India the cow is still revered by Hindu's as a sacred animal being a supporter of life akin to mother nature – the ancient Egyptian goddess Hathor. Hindu's attribute this sacred animal to Lord Krishna who appeared some 5,000 years ago during the Age of Taurus to impart sacred knowledge and wisdom. Lord Krishna and the Archangel Michael are often attributed to one another.

In 2,000 BC, the Age of Aries, witnessed civilizations centered around the symbol of the Ram and a shift from the divine feminine, matriarchal centered society to the patriarchal male dominated society. Hero mythologies and the masculine warrior, fighting and strife, replaced the feminine mother Earth. The prophet Abraham's stories memorialized in history within the Torah are reflective of the patriarchal return.[23] The Ram symbol can be seen in paintings and sculptures depicting the era spoken of as Genesis.

> From Genesis 22:1-18 we are told this truth in a parable, *"Abraham looked up and there in a thicket he saw a ram caught by its horns. He went over and took the ram and sacrificed it as a burnt offering instead of his son. So Abraham called that place 'The LORD Will Provide.' And to this day it is*

[22] http://www.vesicapiscis.co.uk
[23] Ibid.

said, 'On the mountain of the LORD it will be provided.'"

The Age of Pisces began some 2,000 years ago and is seen as the symbol of Christianity. It is the symbol of fertility with circles cutting each other at their centers symbolizing the Vesica Pisces, divine creation and the god/goddess merging of creation. The stories of Mary Magdalene and Jesus as companions fulfilling the Vesica Pisces iconology (stories repressed by the Vatican to this day). In the Eight Beatitudes of Jesus in his Sermon on the Mount Jesus speaks of humility, charity and brotherly love. Similarly under the sign of Pisces are the characteristics of compassion, humility, charity and love. Early Christians used the symbol of the fish (symbol of Pisces) as a secret symbol of their faith. The emphasis on washing of the feet as a ritual signifying purification of the spirit ties into Pisces symbolism as well, for Pisces rules the feet. Pisces "carry" the worries of others and often have sore feet in doing so. Jesus spoke of his role as servant to his flock, which is also a very Pisces notion. Pisces says, "I believe," whereas Aquarius, the age we are in now, says, "Prove it to me scientifically." The Virgin Mary embodied all the qualities represented by the Pisces polarity of Virgo; namely, modesty, commitment to service, and acceptance of what must not be changed. In Pisces, there is a strong need for seclusion, and Christianity puts value on retreats, convents, cloisters or spiritual pilgrimages.[24]

The Age of Aquarius, which we now begin and move through over the next 2,000 years, is the return of God (Prime Creator) astrologically ruled by the planet Uranus known as the "awakener" (causing to "wake up"). It is the age of "Prove it to me scientifically". Aquarius is seen as the return of one with collective social structures and a greater move into the collective

[24] http://www.halexandria.org/dward207.htm

consciousness that births the golden age through spirituality merging with scientific reason. This will be the age where humanity rediscovers its cosmic origins as infinite, quantum consciousness, hence giving rise to the Golden Age – an era of peace and Divine illumination.

Some historical interpretations of the meaning of the Old and New Testament suggest that the Old Testament was a creation of thought from a culture residing within the Age of Aries (ram), and the New Testament is a creation of thought from a culture residing within the Age of Pisces (fish) with stories and parables reflective of the socioeconomic, political and cultural era in which the stories were written. Thus the stories told reflect the iconology of this astrological era.

Similarly, in India the Yuga's (ages) illustrate a cosmic wheel transit every 2,000 years into a new era corresponding to the consciousness on Earth.

A period of “descending epochs” of consciousness:

- Satya Yoga - 12,676 BC
- Treta Yoga (Silver Age) - 9,676 BC
- Dwapara Yoga (Bronze Age) - 6,676 BC
- Kali Yoga (Iron Age) - 3,676 BC

A period of “ascending epochs” of consciousness:

- Ascending Kali Yoga (Iron Age) - 676 BC – 2025 AD, which returns the cycle to the Ascending Golden Age
- **2025 for 2,000 years marks the return of the Golden Age**

The Hindu (Yoga) belief is that within every cycle the Souls who are the actors on that world-drama stage (within that cycle) will be the same. Like a tape recorder playing a song, they believe the Soul repeats its music when the song is played and either descends or ascends. The Soul has a chance to either repeat its karmic past or ascend out of it. Thus if you go back to a period of time, such as the Hebrew exodus, it was the Descending Kali Yoga (Iron Age).

We are now within the ascending Kali Yoga (Iron Age) until 2025. Many spiritualists knowing and following this tradition believe that we have a 20-year transition period from the year 2005-2025 until the Golden Age is realized as a collective consciousness experience and the totality of the return of God as Light (Photonic Consciousness) is understood.

> *"At present, we are undergoing through a critical phase in the history of mankind. This is the period of the confluence of the ending phase of the Iron Age and the starting phase of the Golden Age. This is the most important of all epochs, called the Confluence Age, when God, the Highest Being, descends in this world [as light and vibration] to meet the human beings, His beloved children and gives the most precious boons of Redemption and Beatitude. Through the Godly Knowledge and the easy Raj Yoga, God creates the Golden Age or new vice-less order. "*[25] *Unknown Yogi*

Two-faced Buddha and the Vesica Pisces

In 2013 I took a photo that represents the sacred geometry of the Vesica Pisces as illustrated by the "Two-faced Buddha". The "Two-face Buddha" represents the two aspects of the Buddha's awakening; the "what" and the "how". The Buddha's awakening to pure bliss, love and happiness can be

[25] Quote form unknown Yogi

attained by human effort. As Buddhists will tell you, anyone of us can achieve enlightenment and be the awakened Buddha.

As the Buddha described the awakening experience in one of his discourses, enlightenment is attained through knowledge. The knowledge stages are; knowledge of your own previous lifetimes, knowledge of the passing away, the knowledge of the rebirth of all living beings, and finally insight into the four Noble Truths.

The "Four Noble Truths":

- *The first Noble Truth will be understood by fully understanding Suffering: such was the vision, insight, wisdom, knowing and light that arose in me about things not heard before.*
- *The second Noble Truth of the Origin of Suffering: such was the vision, insight, wisdom, knowing and light that arose in me about things not heard before.*
- *The third Noble Truth of the Cessation of Suffering: such was the vision, insight, wisdom, knowing and light that arose in me about things not heard before.*
- *The fourth Noble Truth of the Path leading to the Cessation of Suffering: such was the vision, insight, wisdom, knowing and light that arose in me about things not heard before....This Noble Truth must be penetrated to by cultivating the Path....right view, right intention, right mindfulness, right concentration….*[26]

[26] http://www.buddhanet.net/4noble.htm

The Buddha who was born as Siddhartha Gautama, lived circa 566 to 480 BC. He was the son of an Indian warrior king who initially led an extravagant life until one day he went in search of understanding and truth. In his search the Buddha realized that to live was to suffer, and when lacking in truth and wisdom the suffering was great. Thus, to be free of worldly suffering one must attain enlightenment through wisdom and truth, and be true in your behavior, thought and intention. The Buddha's four noble truths which were made simplistic for all to understand, teach humanity the truth inherent in life (suffering), the truth of the cause of suffering, the truth of the end of suffering, and finally the path that leads to the end of suffering. This is not a negative perspective of the world, but rather a pragmatic way to acknowledge the challenges inherent in life and how to rise above these challenges to attain joy, happiness and peace irrespective of materialism. In every aspect the Buddha knew that truth, wisdom and light were the keys to joy and enlightenment.

As such, the sphere of light illustrated a profound spiritual truth to me in its representation of the "Two-faced Buddha". This enabled me to discover another aspect of 'knowing' that supported my awakening. On page sixty-one is the photograph I took with the sphere appearing at my elliptical after I had meditated. Inside the sphere is a representation of the Vesica Pisces (overlapping circles), and the "Two-faced

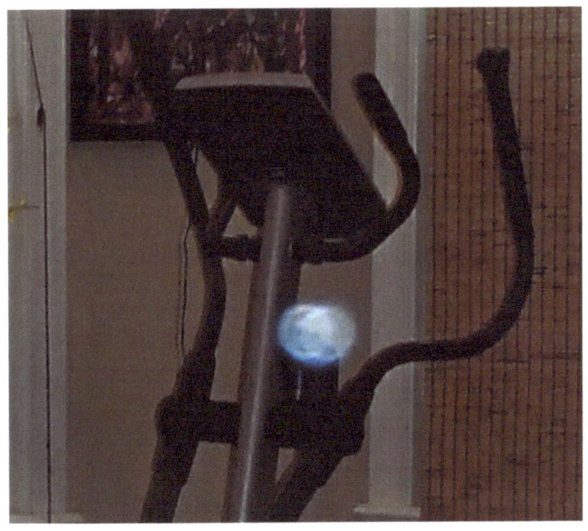

Buddha" illustrating the "what" and "how" of consciousness awakening. Below is a sculpture of the "Two-faced Buddha" for comparison to the sphere appearing in the photograph.

"Emerald Tablets of Thoth"

Thoth, the scribe of the gods, was believed to have been a survivor of a lost civilization that met its demise some 12,500 years ago in a cataclysmic flood and brought sacred wisdom and knowledge to help establish ancient Egypt. Most notably, Thoth is given credit for building the Great Pyramid of Giza and Sphinx. An ancient language was inscribed on a series of Emerald tablets that are thousands of years old. These tablets were translated and made public in the recent past as a precursor to the coming Golden Age. As you will read in the passage below, Thoth speaks of the power of light and the path to consciousness ascension. I recommend you read: *The Emerald Tablets of Thoth the Atlantean* by M. Doreal.

> *"Man has to strive to become the "Divine Sun". Follow this Path and you shall be One with the All. Light comes only to him who strives. Hard is the Pathway that leads to the Wisdom; hard is the Pathway that leads to the Light. Many shall you*

find the stones in your pathway: many the mountains to climb toward the Light. Man, know that always beside you walk the Messengers of Light. Open to all is Their Pathway, to all who are ready to walk into the Light... Know that many dark shadows shall fall on your light, striving to quench with the shadows of darkness the light of the soul that strives to be free. Many the pitfalls that lie on this Way. Seek ever to gain Greater Wisdom. Cognize — you shall know the Light! Light is eternal and darkness is fleeting. Know ever that as Light fills your being, darkness for you shall soon disappear. Open the soul to the Messengers of Light! Let Them enter and fill you with Light. Keep ever your face to this Goal. ... Open the soul, O man, to the cosmos and let it "flow" through you as one with the soul. ...When shining with Soul-Flame, find you the Wisdom, then shall you be born again as Light, and then shall you become the "Divine Sun"." Emerald Tablets of Thoth[27]

Pistis Sophia

From Jesus, the Christ in *Pistis Sophia*, an ancient scroll found in 1773 believing to have come from Egypt in the 2nd Century AD, we read about Jesus' conversation with Mary Magdalene on Mount Olives about "Melchisedec the Great Receiver of Light", meaning "Melchizedek the Over-self Body of Light – the I AM Presence". Lord Melchizedek is also considered a being – a Logos – who is a teacher of the Great White Lodge (a

[27] Thoth, the Atlantean. 1993. *The Emerald Tablets of Thoth-the-Atlantean*"; translations and intrepration by Doreal

reference to light) in which stems the concepts set forth in the Emerald Tablets as elucidated in the quote on page eighty.

> *"It came to pass then, when Jesus had finished saying these words unto his disciples, that Mary [Magdalene], the fair in her discourse and the blessed one, came forward, fell at the feet of Jesus and said: "My Lord, suffer me that I speak before thee, and be not wroth with me, if oft I give thee trouble questioning thee."*
>
> *The Saviour, full of compassion, answered and said unto Mary: "Speak the word which thou willest, and I will reveal it to thee in all openness."*
>
> *Mary answered and said unto Jesus: "My Lord, in what way will the souls have delayed themselves here outside, and in what type will they be quickly purified?" And Jesus answered and said unto Mary: "Well said, Mary; thou questionest finely with thy excellent question, and thou throwest light on all things with surety and precision. Now, therefore, from now on will I hide nothing from you, but I will reveal unto you all things with surety and openness. Hearken then, Mary, and give ear, all ye disciples: Before I made proclamation to all the rulers of the æons and to all the rulers of the Fate and Of the sphere, they were all bound in their bonds and in their spheres and in their seals, as Yew, the Overseer of the Light, had bound them from the beginning; and every one of them remained in his order, and every one journeyed according to his course, as Yew, the Overseer of the Light, had established them. And when the time of the number of Melchisedec, the great*

Receiver of the Light came, he was wont to come into the midst of the æons and of all the rulers who are bound in the sphere and in the Fate, and he carried away the purification of the light from all the rulers of the æons and from all the rulers of the Fate and from those of the sphere--for he carried away then that which brings them into agitation--and he set in motion the hastener who is over them, and made them turn their circles swiftly, and he [sc. the hastener] carried away their power which was in them and the breath of their mouth and the tears [lit. waters] of their eyes and the sweat of their bodies.

"And Melchisedec, the Receiver of the Light; purifieth those powers and carrieth their light into the Treasury of the Light, while the servitors of all the rulers gather together all matter from them all; and the servitors of all the rulers of the Fate and the servitors of the sphere which is below the æons, take it and fashion it into souls of men and cattle and reptiles and wild-beasts and birds, and send them down into the world of mankind. And further the receivers of the sun and the receivers of the moon, if they look above and see the configurations of the paths of the æons and the configurations of the Fate and those of the sphere, then they take from them the light-power; and the receivers of the sun get it ready and deposit it, until they hand it over to the receivers of Melchisedec, the Light-purifier. ... According to the circle of the rulers of that sphere and according to all the configurations of its revolution, and they cast them into this world of mankind, and they

become souls in this region, as I have just said unto you."

"According to the circle of the rulers of that sphere and according to all the configurations of its revolutions" [Meaning in accordance to the Soul's evolution in consciousness it is cast into the world of mankind.] "The Light Purifier, the I Am presence, decides the incarnates fate based upon is 'revolutions'."

"It came to pass then, when those tyrants saw the great light which was about me, that the great Adamas, the Tyrant, and all the tyrants of the twelve æons [meaning 'reptilian' human brain of physical matter and the 12 houses of illusion], all together began to fight against the light of my vesture, desiring to hold it fast among them, in order to delay in their rulership. This then they did, not knowing against whom they fought. "When then they mutinied and fought against the light, thereon by command of the First Mystery I changed the paths and the courses of their æons and the paths of their Fate and of their sphere. I made them face six months towards the triangles on the left and towards the squares and towards those in their aspect and towards their octagons, just as they had formerly been. But their manner of turning, or facing, I changed to another order, and made them other six months face towards the works of their influences in the squares on the right and in their triangles and in those in their aspect and in their octagons. ... "But when I had taken away a third of their power, I changed their

spheres, so that they spend a time facing to the left and another time facing to the right. I have changed their whole path and their whole course, and I have made the path of their course to hurry, so that they may be quickly purified and raised up quickly. And I have shortened their circles, and made their path more speedy, and it will be exceedingly hurried. [the Singularity]

"And they were thrown into confusion in their path, and from then on were no more able to devour the matter of the refuse of the purification of their light. And moreover I have shortened their times and their periods, so that the perfect number of souls who shall receive the mysteries and be in the Treasury of the Light, shall be quickly completed. ...

"For these powers have been given unto you [Mary Magdalene] before the whole world, because ye are they who will save the whole world, and that ye may be able to endure the threat of the rulers of the world and the pains of the world and its dangers and all its persecutions, which the rulers of the height will bring upon you. For many times have I said unto you that I have brought the power in you out of the twelve saviours who are in the Treasury of the Light. For which cause I have said unto you indeed from the beginning that ye are not of the world. I also am not of it. For all men who are in the world have gotten their souls out of [the power of] the rulers of the æons. But the power which is in you is from me; your souls belong to the height.

I have brought twelve powers of the twelve

> *saviours of the Treasury of the Light, taking them out of the portion of my power which 13. I did first receive.*[28]

The above ancient text, *Pistis Sophia*, may seem off topic for this chapter, but as you read between the lines to grasp the depth of meaning, you can understand in this magnificent teaching given by Jesus to Mary Magdalene and his Apostle's the importance of the message. In the discourse Jesus foretells of humanities consciousness evolution on Earth, its descent from and ascension back into the Treasury of Light, and the cosmic wheel of ages that foretells the 'time' for humanity to prepare for the cosmic awakening – the return of God. I highly recommend you read *Pistis Sophia*, Dr. J.J. Hurtak's translation version. It provides extraordinary insight into the magnificent wisdom and teachings of Jesus who held the cosmic "Law of One" in his body, mind and Soul as he walked on Earth. A great gift to humanity.

[28] Hurtak, J.J. PhD. 1999. *Pistis Sophia*: A Coptic Gnostic Text with Commentary

CHAPTER SIX

HARMONICS AND THE PHOTONIC WAVE

"My soul is a hidden orchestra; I know not what instruments, what fiddle strings and harps, drums and tamboura I sound and clash inside myself. All I hear is the symphony."
Fernando Pessoa, The Book of Disquiet

The harmonic nature of creation is a marvel to explore within the quantum information field. As we know, atoms give rise to matter and each atom moves at its own speed, however when combined with other units they create a specific vibration in the atomic field. This motion produces pressure and this creates the harmonic waves (sound) of creation. The multiple waves created by the vibration of atoms shift the pitch of the harmonic wave like various instruments in a symphony uniting in a crescendo of sound.

"In the beginning was the word, and the word was with God, and the word was God."
John 1:1

It is the photon acting as a wave particle singing and dancing its way through all dimensions of creation that give rise to the quantum information field of light, or as you read in the words of Jesus in his discourse in *Pistis Sophia*, "Treasury of Light".[29]

The harmonic wave created by the movement and dance of the photon affects everything in all dimensions. It is the cosmic beating of God's heart, love is its frequency, and sound is its song – the "word". The pulsed cosmic heartbeat of God is synchronized to all of creation and human consciousness.

The Orchestra of Light

Photons switch on the body and mind processes like an orchestra conductor bringing each individual instrument into the collective sound. At different frequencies, they perform different functions. For instance, the molecule having its own unique electromagnetic (EM) field somehow "senses" the EM field of a complimentary molecule. A dance ensues between the cellular medium and molecules, which move to the wave rhythm of the frequency (pitch) by which they vibrate, and the biophoton conducts the music in the third dimensional plane.

The scientific methodology that examines wave phenomenon and vibration, and the transformational nature of sound is called Cymatics. Endless experiments have illustrated the distinct similarity between the standing wave patterns of resonance and the geometrical form of plants and animals, which in turn reflects the mathematical sequence found in the Golden Ratio of Phi and the Fibonacci sequence.

In his book, "Water Sound Images", Alexander Lauterwasser created montages of these

[29] Hurtak, J.J. PhD. 1999. *Pistis Sophia*: A Coptic Gnostic Text with Commentary

ascending harmonic forms of sand on steel plates, and in water… As the water is subjected to gradually increasing frequencies, the complexity of the patterns increases with the pitch of the exciting tone. At a critical pitch, the sample dissolves into chaos, only to re-configure into a higher order of complexity as the tone continues to ascend. This process of chaos and re-integration continues as the frequency ascends, while the periods of chaos become shorter in duration and arise at more frequent intervals.

This process may be observed in many phenomena, from the valence fields of electrons within the atom (whose complexity increases as you ascend the Periodic Table of Elements), to the harmonic series that is generated each time a string is plucked, to complex turbulences that create weather patterns, to the intricate physiological processes within our bodies that allow us to function in homeostasis within an ever-changing sea of vibrations (our environment). I like to think of it as a "living metaphor" for the process of evolution that occurs at all levels of creation, from the physical to the subtlest domains of consciousness, where its implications are even more impressive.[30]

Sound and the Multidimensional Universe

Think of the multidimensional universe as light/sound frequency within the subtle energy of photonic consciousness transiting in our dense universe as biophotons.

Sound is the key driving force behind manifestation of form and matter through the pitch of its resonance. If you are

[30] Volk, Jeff. 2010. *From vibration to manifestation: Assuming our rightful place in creation*: 9-10

tuned into that vibration, the harmonic frequency is heard and understood by your body, mind and Soul. Your ability to reside within a frequency band of photonic consciousness corresponds to the depth by which you resonate with the corresponding signal. And, by focusing your conscious intention on light (truth, wisdom and love from God), you absorb more photonic consciousness at the molecular and atomic level and thus ascend in body, mind and soul.

Subtle energy within the quantum information field is eternal and infinite in form and expression. The photon is the means by which consciousness is expressed, and sound is the means by which consciousness solidifies into physical form, as light and sound is the essence of creative forces in the Kingdom of God.

It is the God Consciousness in everything that directs the intricate dance, the music of the atom that gives rise to the physical universe, and our Soul that animates matter and brings it to life. This is the cosmic dance of matter and life. From ancient texts we can intuit that higher dimensional beings we think of as Archangels understand this divine law and direct this process through the vibration of love, balance and harmony.

As you allow for the increased absorption of more light – photonic consciousness – your frequency rises and you are tuned into the channel of divine love, God's channel. I believe that the language of God is love, and we as "frozen light" residing in the third dimension are transmitters and receivers of frequency.

For example, think of a radio. You set your dial to what you choose to listen to. Thus the simplicity of divine creation is this; you are a co-creator of your reality basking in the conscious, eternal and infinite frequency of Prime Creator, God. You have a choice in setting your dial. I choose to set my dial to God's channel, what do you choose?

The Science of Harmonics

Modern physicists have not yet identified the primary source of the photonic and acoustic energies arising and distinct to living and inanimate matter. Vibratory relationships observed among the atoms comprising our bodies and all matter are recognized as the essential key for understanding elemental nuclear transmutations, which opens the door to a harmonious technological era of atomic resonance, with atomic meaning "atom".

The symphony of atoms in a crystal lattice generates the dynamic relationships that facilitate the natural energy conversion of one element into another. Many ancient cultures believed in the physical connection to cosmic harmonic resonance. For instance, when we examine my photographs of the "sphere's of light", we observe the fluid wave-like nature of photonic consciousness visibly observed as a biophoton. Although the spheres may look small to us, it is in fact a magnificent universe in unto itself.

These spheres are much bigger and grander than our perceived world, and we in this lower dimension are truly swimming in the "fish bowl". We are the little fish just figuring that out.

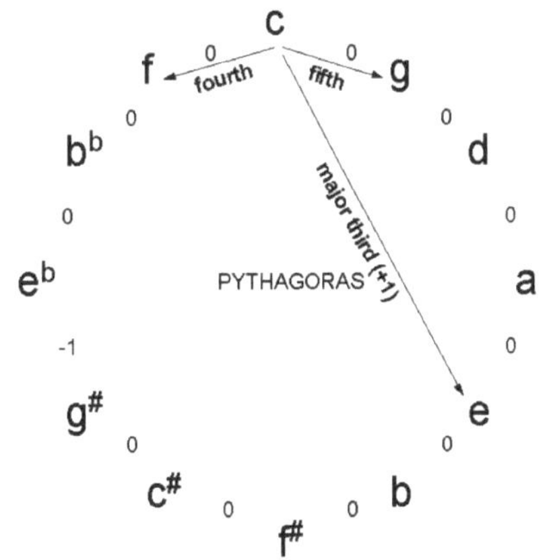

The circle can be understood as an illustration of divine consciousness and its infinite, eternal existence – the Alpha and Omega. It cannot be destroyed. Just as energy cannot be destroyed, it transforms and changes form.

Along the points of the circle are infinitely vibrating rates on an infinite scale of possibilities, constantly moving up and down in octaves – sound resonance scales. In the graphic above is an illustration of the circle beginning with the octave "C". This is the "Pythagorean temperament", a full chromatic scale created by perfect pitch.[31] Notice in the graphic to the right the spiral of sound ascending from the octave scale, the "C" as it rises in pitch.

Musicians and physicists recognize pitch as being multidimensional.

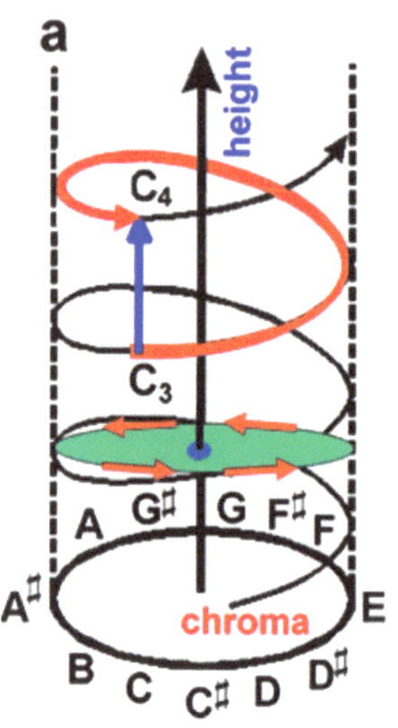

This is referred to as "pitch height" or "pitch chroma". The graph above illustrates the "helix circle sound" that ascends up the chroma dimension around the pitch helix.[32] Chroma is the quality of pitch, also defined as purity of color.

"C" begins the major scale of a pitch class assigned the number "0" and "1". Pitch classes are 0-11, and "C" major scale is the most common key in music. This gives a new and profound meaning to the "C" that appeared in the photo I took of the sphere, and in the crop circle illustrated in chapter three.

---

[31] http://hyperphysics.phy-astr.gsu.edu/hbase/music/pythag.html

[32] Patterson, R.D. 1990. *The tone height of multi-harmonic sounds.* 8, 203-214 http://www.pdn.cam.ac.uk/groups/cnbh/teaching/sounds_movies/pitch_helix.htm

"C"

> *"Today there is a wide measure of agreement, which on the physical side of science approaches almost to unanimity, that the stream of knowledge is heading towards a non-mechanical reality; the universe begins to look more like a great thought than like a great machine. Mind no longer appears as an accidental intruder into the realm of matter; we are beginning to suspect that we ought rather to hail it as a creator and governor of the realm of matter...* Physicist, Astronomer & Mathematician James Hopwood Jeans[33]

All living matter is comprised of four fundamental elements, which are the building blocks of life: hydrogen, nitrogen, oxygen and carbon. Carbon ("C") comprises almost all of the most intricate structures of matter, and it comes from our cosmos. Carbon transits across the Milky Way galaxy from the central sun to our solar system, and sun. Our sun then disperses the building blocks of life throughout our solar system via the solar winds, something I will discuss in detail in the Stargate chapter.

Carbon "C" atoms enter the terrestrial biosphere of Earth from the cosmos and is synthesized, metabolized, by all living matter. For plants, this process is called photosynthesis, whereby plants and other living organisms convert the photon (light) into chemical energy. We humans then consume plants, and animals which consume plants, and metabolize this light within our own physical body. We are literally metabolizing

[33] Jeans, James. 1930. The Mysterious Universe: 137

quantum information encoded within light.

The charged carbon atoms streaming from the gamma-ray bubble in the center of the Milky Way galaxy are penetrating our solar system and affecting an ascension at the molecular level. This dance of light is also the song of the ascension, and is currently being recorded by NASA and scientists studying the auditory composition of this high frequency light.

It is interesting to note that the carbon ion has a frequency hum that directly correlates to the speed, fast or slow, by which the solar winds move. The higher the frequency, the faster it moves, and this sound creates a sonic "boom" at it penetrates our terrestrial biosphere. This can be understood as the Doppler Effect, which is a shift in the frequency sound wave (high or low) produced by movement - speed.

Could this be the source of the sonic booms and trumpet like noises being heard all around the world for the last several years? Could the carbon atom, "C", have been known and understood thousands of years ago, and explained in ancient writings, like the *Book of Revelations*, as a cosmology to identify the coming ascension?

Scientists will tell you that it wasn't until the 17th Century that carbon was discovered as an element, and in the 19th Century that the Doppler Effect was discovered by Austrian physicist Christian Johann Doppler, however there exists a more ancient understanding and importance of the "C" within the cosmic wheel of life.

In the *Book of Revelation*, the number 666 is used to identify the 'beast'. In an attempt to control the behavior of man, by instilling fear, this phrase is interpreted to identify the nemesis of mankind, Satan – the beast, however I believe it is a literal statement pointing humanity to the great truth of the carbon atom, in which living matter cannot exist without.

From the King James Bible: "Here is wisdom. Let him that hath understanding count the number of the beast: for it is the number of a man; and his number is six hundred threescore and six" Revelation 13:18

The carbon atom contains 6 electrons, 6 neutrons and 6 atoms, "666", and in the Periodic Table of Elements, the atomic number of carbon is "6. The cosmological reference made in Revelations 13:18 could not be more clear.

"And the seven angels who had the seven trumpets prepared themselves to sound them." Revelation 8:6

The world shift into the Christ Consciousness – Golden Age – is described in the *Book of Revelations* with vast detail. A new interpretation of the Seven Seals illustrates a cosmology, a scientific understanding of this shift, as well as a spiritual aspect through the awakening of the seven chakras (seals) within the human body. For more on my interpretation of the Seven Seals please read my book *Approaching Singularity: The Genesis of Creation*.

Physical Universe of Matter as Light and Sound

Think of dimensions as sound frequency with light as an energetic means by which eternal divine consciousness travels. If you are tuned into that vibration, the frequency (sound) is heard, and understood by you and your 'light', or consciousness, then resides in it. It moves, it vibrates and it is constantly changing in form. In the physical world of matter, where life is temporary and based on entropy, light becomes frozen creating the appearance of the solidity of matter.

Our physical body is a bio-vessel for our conscious, Soul

expression in this dimension. Our eternal Soul must have a vessel to walk about this physical world to enable us to experience energy in motion (emotion/feeling) and be a co-creator of matter through our biological and mental creative processes.

Our physical universe of matter existing within the space-time continuum is ordered on a shortwave frequency of light and sound. Sound moves in waves, and in turn creates fields. It is the nonlinear interaction of these short and long-waves that produce patterns of harmonic resonance throughout our universe. Our planetary harmonic resonance is synchronized with human consciousness, the beating of all hearts and the alpha rhythm of the human brain.

Studies show that when multiple systems of the body are vibrating together in the frequency of positive, higher vibrational light; such as, love and joy, the overall physical and mental health of the human is good. Conversely the reverse happens with negative thoughts resonating with lower frequency light. Everyone, just like our planets and universe, generate their own harmonic vibration. We are tuned to be sympathetic to the vibration of Earth, which is referred to as the Schumann resonance.

> *[In the 1950's] German physicist W.O. Schumann of the Technical University of Munich predicted that there are electromagnetic standing waves in the atmosphere, within the cavity formed by the surface of the earth and the ionosphere. This came about by Schumann teaching his students about the physics of electricity. ... In 1954, when measurements by Schumann and König detected resonances at a main frequency of 7.83 Hz.*[34]

[34] http://www.earthbreathing.co.uk/sr.htm

The Schumann resonance illustrates that electromagnetic impulses generate in the spherical earth-ionosphere as electrical discharges, e.g. lightning. As we can observe, lighting exemplifies both light and sound as a powerful force of mother nature. These electrical discharges have both high frequency (short) and low frequency (long) waves. The waves that navigate the earth have a frequency of 7.8 Hz, and it is the waves of this frequency and its harmonics at 14, 20, 26, 33, 39 and 45 Hz that form the Schumann resonance.[35]

The resonance of Earth is not caused by anything internal to Earth. Like all matter, Earth is comprised of vibrating atoms illuminating, dancing and singing into the physical form we experience as Earth. Like us, Earth receives photonic consciousness and subtle energy within the quantum information field, and this gives rise to 'her' frequency and vibration that we call the Schumann resonance. Thus, the resonant heartbeat of Earth is far more than electrical discharges we observe as lightning. Lighting is a manifestation of this process that we can observe.

As scientists have proven, the alpha rhythm of the human brain resonates with the natural frequency of Earth synchronizing life's electromagnetic fields in a symbiotic relationship with one another. This symbiotic relationship enables photonic consciousness that passes through our Sun to Earth to resonate with life on her vessel. This is a fascinating topic in itself that takes you into the 'stargate rabbit hole', a topic I will attempt to expound on in chapter eight.

"Listen, I tell you a mystery: we will not all Sleep, but we will all be Changed — in a flash, in the

[35] Ibid.

twinkling of an eye, at the last trumpet. I Corinthians 15:51

As our solar system enters the photon belt, our resonances are ascending together as more higher vibratory light – photonic consciousness – descends into our reality.

> *"Understand that the Earth plane is now being lifted into a higher vibratory acceleration which means that Earth's solidity (carbon) is now being rearranged so that matter can be changed into silicon (crystal light). Therefore, planet Earth is graduating, and once again joining the higher heavens. As a consequence, all the souls who let go of negativity can become one in divine love and also arise in Earth's great ascension."*[36]

---

[36] Phylos, Orpheus. 1999. *Earth, the Cosmos and You*

CHAPTER SEVEN

DNA

"It is impossible to account for the creation of a Universe without the agency of a Supreme Being."
George Washington

Photonic consciousness is the macrocosmic conduit by which information is transferred into our DNA through absorption and emission at the biophotonic level. Thus in all things physical, it can be understood as frozen light with DNA being the library of this frozen light information. DNA is both the story and song of photonic consciousness frozen in the space-time continuum of the third dimension.

> *Dr. Leonard Horowitz of Harvard University illustrated that the primary function of DNA lies in the realm of bio-acoustic and bioelectric signaling. Horowitz showed that DNA emits and receives photons or electromagnetic waves of sound and light.*[37]

DNA is the code of life housing unique genetic information in the nucleus of every cell. Within the quantum information field often referred to as a "zero-point field", it is the subtle energy

[37] Horowitz, Dr. Leonard. 2004. *DNA: Pirates of the Sacred Spiral Book.*

field of photonic consciousness that brings information into all of creation. As I have explained in this book, the quantum information field is an unseen, subtle energy of photonic consciousness that is all around us. As this multidimensional energy is brought into the physical third density it becomes a biophotonic field of light information that vibrates at different frequencies along the spectrum of light. Thus, our DNA is biophotonic light; e.g. frozen light, and we, like all matter, are biophotonic organisms.

As organisms do emit light, we might presume that human beings residing at the top of the food chain would emit the greatest amount of light; however this is not the case, and the reason for this is a secret locked within our DNA. Humans only emit 10 photon/cm2/sec where as plants and some base animal life (meaning animals without a cerebral cortex, which excludes all mammals and birds) emit the highest at 100 photon/cm2/sec within a high frequency of the electromagnetic spectrum of light. Curious I asked myself the question "why", and the answer to this question leads you to the metaphorical "fountain of youth", with good health and happiness.

A theoretical biophysicists who has done amazing research in this field, F.A. Popp, ultimately concluded that matter is a biophotonic organism, which is surrounded by a quantum information field of light, and that DNA responds and interacts with photonic information in this field.[38]

Biophysicists state that DNA is a bio-communication system and acts as a quantum resonator, efficiently storing the photon by internal reflection of nonlinear standing waves. Popp's research illustrates that DNA is a light storage house that drives the bodies functioning and processes. *"We know now, today, that man is essentially a being of light,"* stated theoretical

---

[38] Dale, Cindy. 2009. *The Subtle Body: An Encyclopedia of your energetic anatomy*: 44

biophysicist Popp.[39]

Popp's studies of DNA have concluded some miraculous understandings of why some cells can become cancerous, and it all has to do with light and the frequency by which it vibrates. In understanding the conscious creative power of light, Popp experimented on plants using photosynthesis. Popp ultimately concluded that the organism, in this case humans, that eats plants ingests the light stored within the plant and in doing so is metabolizing light and this high vibratory photonic energy is distributed over the spectrum of electromagnetic frequencies of the body.[40]

Could this be why God designed plants and base animal life to emit the highest frequency of light? This certainly is a powerful argument for vegetarians and pescatarians, e.g. vegetarians who eat fish and crustaceans, which do not have a highly developed cerebral cortex. As I began my awakening I had the intuitive instinct to make drastic changes to my diet. Without understanding why I became a pescatarian, and in doing so the influx of light exponentially increased my awakening.

As Popp's research proved, the higher the animal on the food chain the less light it emits, and as you will read on with Popp's research, high frequency light absorbed into the body through metabolism directly corresponds to good health. As Popp's experiments illustrated, the introduction of low frequency light metabolized into the body at the subatomic level can breakdown communication between cells that can manifest as illness, most particular cancer. He concluded that good health is achievable when humans balance the infusion of high frequency light into their body.

This can be done both physically and spiritually. Physically we can ingest foods containing more light, i.e. healthy plant food,

[39] http://www.viewzone.com/dnax.html

[40] Ibid.

and spiritually through our thought processes of attracting higher frequency photonic consciousness to us at the atomic/molecular and DNA level. Remember in subtle energy, like attracts like, meaning loving, positive thoughts will attract higher frequency photonic consciousness that is the physical and spiritual healing elixir of God.

DNA the "Master Tuning Fork"

When we think of DNA we can intuit it as a metaphorical "Master Tuning Fork" of the physical body."[41] DNA is tuned to a specific frequency and the cells follow the instructions of this information. Some theoretical biophysicists believe this is the missing link that explains why a single cell can turn into a fully formed organism, such as human.

Geneticists have also found that DNA retains far more than genetic, light information, but also retains a kind of energy information they describe as emotional or behavioral. Geneticists have discovered that DNA is changing throughout the life of the human and does not remain stagnate. Thus DNA infused with biophotonic information will instruct the cells accordingly. In other words, high frequency photonic consciousness and biophotonic infusion into the mind and body will support a healthy quality of life. For instance, geneticists have scientifically found that the lack of love can turn off the genes that support the healthy functioning of the hippocampus in the brain, which is important for memory and emotional responses. I propose that it is the combination of high frequency light that is metabolized into the body and cellular structure from the physical world of plants, and cosmic world of photonic consciousness, that is the elixir to good health, happiness and consciousness ascension.

---

[41] http://www.viewzone.com/dnax.html

CHAPTER EIGHT

STARGATE

"I come from the Central Sun beyond your Universe, or as you understand it, from the seat of God, using the power of breath that extends beyond time into timelessness. My breath of energy is of the size of a galaxy permeating through many dimensions...."
The Archangel Michael speaks[42]

Wormholes have been the stuff of science fiction since being predicted by Albert Einstein. In 1935, Albert Einstein and fellow physicist Nathan Rosen proposed the theoretical existence of "bridges" through space-time. These bridges are known as "Einstein-Rosen Bridges" or wormholes. Science fiction enthusiasts call them "stargates", and I prefer to use this word as it simply more fun.

Stargates are bridges through space-time creating shortcuts between universes. The theory suggests that a wormhole could connect two different points in space-time enabling a traveler to pass between distant universes. Since Einstein - Rosen theorized the possibility of this extraordinary concept NASA has discovered that wormholes (stargates) do indeed exist naturally on Earth. One could theorize that if stargates exist on Earth, than they would exist throughout the

[42] Phylos, Orpheus. 1999. *Earth, the Cosmos and You*: 1

cosmos and perhaps be activated in our own electromagnetic field.

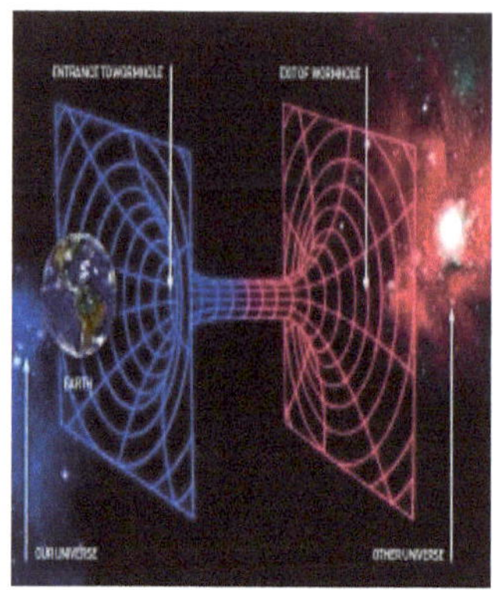

Quantum physicists studying multi-dimensional planes of existence and subtle energy, theorize that wormholes could create pathways between different dimensions. In theory, a wormhole could create a pathway between the third dimension, where we physically reside, and the fifth dimension existing outside the space-time continuum, e.g. a multidimensional stargate.

How would these wormholes appear, you might ask? Physicists theorize that wormholes, which occur naturally, would contain two openings like mouths and be connected by a throat between points. A graphic illustration is above.

This theory suggests that the openings would look like a swirling sphere of negative mass, and that wormholes may contain a kind of particle scientists don't understand called, "exotic matter". Exotic matter exists outside the boundaries of gravity, known as antigravity, and has exotic properties.

Exotic matter is defined as hypothetical particles, which have "exotic" physical properties that would violate known laws of physics, by having negative mass. Negative mass particles are repelled by gravity and accelerate in the direction opposite of applied force. For example negative mass is attracted to like charge; meaning negative mass with a positive charge is attracted to mass with a positive charge – <u>like attracts like</u>. Sounds like a variation of "subtle energy", which defies the laws of Newtonian physics and physical matter wherein opposites attract. Scientists have a hard time reconciling negative mass and exotic matter as it violates the laws of conservation of momentum or energy.

Quantum physics state that exotic matter travels faster than the speed of light. As I have documented in the over 4,000 photographs I've taken over the years, photonic consciousness transiting through the quantum information field of subtle energy does in fact exist. For myself, family, and close friends who have witnessed this phenomenon, this is an observed reality.

As photonic consciousness is real, could stargates actually exist and be the bridge by which photonic consciousness passes from higher dimensional planes outside of the space-time continuum into the third dimension? Is this a type of subtle energy, multidimensional intergalactic 'highway'?

Do these stargates exist solely for the purpose of enabling creative energy coming from higher dimensions of the quantum information field to enter this physical dimension for the purposes of encoding and directing creation?

In short, I believe the simple answer is "Yes!" Photonic Consciousness descends from higher dimensional planes through the "central sun" at the center of the Milky Way galaxy, which acts as a doorway between dimensions, and then transits through a series of stargates within each solar system – an intergalactic highway of light. This intergalactic highway of light is similar to the neural network within the physical body – "As above, so below…"

I may not be able to scientifically prove my theory about the purpose of the intergalactic highway of stargates yet, but I believe that a primary purpose of our Sun is to receive creative conscious energy that transits through its innate 'bio-stargates', and deliver photonic consciousness in the form of biophotons to Earth. Earth with its own innate 'bio-stargate' then metabolizes this light frequency that in turn is metabolized by all life resonating on her vessel.

Sound crazy? Well as you will read, natural bio-stargates do exist and NASA is using taxpayer dollars to fund research to understand the nature of the wormholes that connect Earth to our Sun. Maybe it is not such a crazy theory after all, and in fact if proven true, it is a wonderful illustration of the magnificent achievement of cosmic engineering that is our universe.

It reminds me of the human brain transmitting neural signals throughout the body telling it what functions to perform.

Natural Stargates

As I previously mentioned, NASA has theorized that wormholes naturally exist around Earth and are known as "X-points". The theory suggests that Earth's electromagnetic field mingling with solar winds and the electromagnetic field of the Sun connect uninterrupted pathways that warp space-time to create wormholes.

NASA announced the discovery of these hidden portals in Earth's magnetic field referring to them as electron diffusion regions. Electron diffusion regions are areas where the electromagnetic fields of the Earth and Sun create a bridge to our Sun some 93 million miles away.

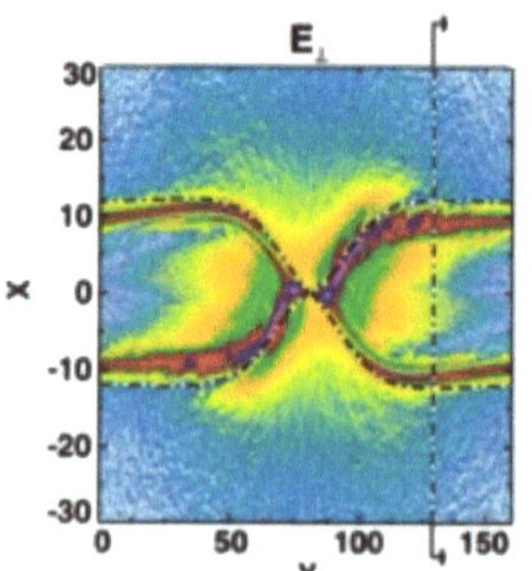

Data from NASA's Polar spacecraft, circa 1998, provided cruicial clues to finding magnetic X-points. Image Crdit: NASA

NASA has used both the THEMIS spacecraft and European Cluster probe to examine this phenomenon and has subsequently reported their findings suggesting that these X-points open and close dozens of times each day, and they don't yet understand how or why. Scientists believe that these X-points enable the transfer of the magnetic field and exotic energy (subtle energy) from the sun to earth.

I theorize that this exists as a kind-of neural network of information transfer in the form of higher vibrational photonic consciousness resembling the electrical transfer of information from the brain throughout the body. In this case our Milky Way galaxy can be seen as the macrocosmic body of this universe.

NASA is so interested in this phenomenon that it has launched a new mission this year called Magnetospheric Multiscale Mission (MMS) as a Solar Terrestrial Probe mission comprising four spacecraft that will study Earth's magnetosphere analyzing microphysics for magnetic reconnection, energetic particle acceleration and turbulence.[43] NASA claims its intent is to better understand the microphysics of space weather, and University of Iowa professor contracted to support the project, Dr. Scudder, acknowledges this is about understanding magnetic portals, i.e. X-point - wormholes. Dr. Scudder of the University of Iowa states, *"Magnetic portals are invisible, unstable and elusive, they open and close without warning and there are no signposts to guide us in."*[44]

An Electric Cosmology

There is a growing community of scientists and researchers who believe our universe is electric, and is a constantly morphing organism in which all its integral parts affects the whole, similar to the physical body of any living organism. For example, the electric currents that power the galaxies flow within its magnificent spiral arms.

This is reflective of the same electric currents flowing in the human body. This electrically charged plasma, known as solar winds, are fast moving protons that carry an electric charge, and as it passes through the human body creates a

---

43 NASA official website; http://mms.gsfc.nasa.gov

44 NASA official website; http://www.nasa.gov/mission_pages/sunearth/news/mag-portals.html#.U2u-nP3nf1o

magnetic field that induces an electrical current beneficial for our neurological body functions.

The central sun in the center of our Milky Way galaxy occupies the highest plasma rich space in our galaxy, and it is theorized within the electric universe model that our Sun was formed by concentrated plasma streams coming from the center of our galaxy. This theory suggests that the central sun gave birth to our solar system, and in fact all star systems in our galaxy. This correlates to the Mayan belief of the "Hunab Ku" being the "womb of creation, giving measure and motion" to our universe.

Proponents of the Electric Universe model believe that the creative process for the solar system is illustrated in the distribution of atomic elements across the planets that flow from our Sun. Could this prove we live in an electrically powered universe in which the Sun is the distribution mechanism with the solar winds carrying the elements by attraction? This would mean that each planet in our solar system attracts elements specific to its composition.

This reminds me of the functioning of the human body, whereby each organ has a specific function and attracts specific elements for the performance of the whole. For instance, our bodies consist of cells that are organized into many specialized organs and tissues to perform a variety of functions. Our stomachs digest food so that the nutrients contained in the food can be distributed to the rest of the body. Our lungs take in the oxygen needed by the body and release carbon dioxide as a waste product. Our muscles allow the body to move. Our brains coordinate all of these (and many other) activities of the body. These processes are based upon many different chemical reactions, and the sum total of the chemical reactions in the body is known as the body's metabolism.

I propose that on a grander cosmic level, our Sun is the

metabolism of our solar system, with the solar winds as the breath of God delivering the elements of creation comprising matter ("c").

The chemical reactions happening on the surface of our electrically charged sun, do comprise the sum total of atomic elements dispersed throughout our solar system making up our planets and maintaining their vital functions. The Milky Way galaxy, our solar system, Sun, Earth and all planets therein, are an integrated whole.

Similarly we know that the various components of our body are an integrated whole. And, in a grader infinite perspective, our body (matter), mind (thought), and Soul (Photonic Consciousness) are the greater integrated whole. "As above, so below..."

The Photonic Wave

The universe is held together by the electric force of the Photonic Wave. It is known that not a single atom would exist without the electric force being expressed, and our solar system would not exist without the atomic electric force. Thus, planets are formed by atomic elements that have been metabolized on the surface of the Sun, and distributed throughout the solar system within the solar winds (i.e. plasma events) within the "X-Point" wormhole gateways – the intergalactic highway.

It is the convergence of large plasma events, like Super Nova's, that create black holes. Science do not yet understand the function of a black hole, however it is known that black holes are essentially massive high frequency photonic (light) events.

I propose black holes are massive high frequency photonic events that create doorways to higher dimensional planes of existence, such as the 5th dimension and beyond. Thus, a Super Nova event is an electrically charged photonic event that births gateways for evolution of the universe through the influx of higher frequency light, and quantum information. These gateways cannot be seen, but are subtle energy connecting this reality to higher dimensional planes.

Although scientists currently perceive a black hole to be a destructive force, I believe it is a magnetically neutral zone that is the gateway (doorway) between dimensions. Returning back to the gamma ray bubble extending from the center of the Milky Way galaxy, we can intuit that this intergalactic plasma stream(s) feed our galaxy visa vie suns, and its currents flow within a intergalactic highway of light.

Evidence of Stargates

As I explained in the chapter on subtle energy, it is massless and exists as an unseen energetic creative force that can be observed in the ultraviolet spectrum of light as a biophoton. Although our eyes cannot see in the ultraviolet spectrum of light, various camera technology and some animals can. Cameras with a broader observational field of the spectrum of light are commonly used in scientific study. This helps to explain the phenomenon of capturing "orbs or plasma spheres" in photographs.

On page ninety-two is a picture I took of a stargate that opened at my elliptical machine in 2013.

You may be wondering why so much 'action' occurs at my elliptical machine? This is an important question that answers fundamental aspects of attracting the subtle energy of photonic consciousness into your electromagnetic field.

In my bedroom where my elliptical machine is located also resides a quiet space where I meditate and pray. My meditations are always done with positive, loving conscious intention and focused on the absorption of high frequency light from God. My elliptical machine is a place where I practice a different kind of meditation. As I exercise, and enter into a rhythmic physical movement, I use sound and conscious intention for mind clarity. It is like entering the "runner's trance". I will explain more about this in the next chapter on Super Consciousness.

As such, conscious intention combined with your energy from your electromagnetic field brings forth miraculous phenomenon, such as the appearance and opening of wormholes that are excited, or energized. This is done through the raising of your vibration – frequency. It is a microcosmic example of the super consciousness innate in humans, an innate function of planets such as Earth, which is now being scientifically studied.

In the photograph on the next page, are pictures I took within seconds of each other that illustrate the acceleration and opening of a wormhole. On the left is the original photo and on the right is the cropped/zoomed perspective for closer examination of the wormhole opening and expanding. As you will notice from the first photograph, to the second photograph, the wormhole opens in a deepening progression and moves slightly to the right of the elliptical. In the second photograph, you can see the deepening progression of the wormhole into a visible three dimensional funnel.

First photograph below.

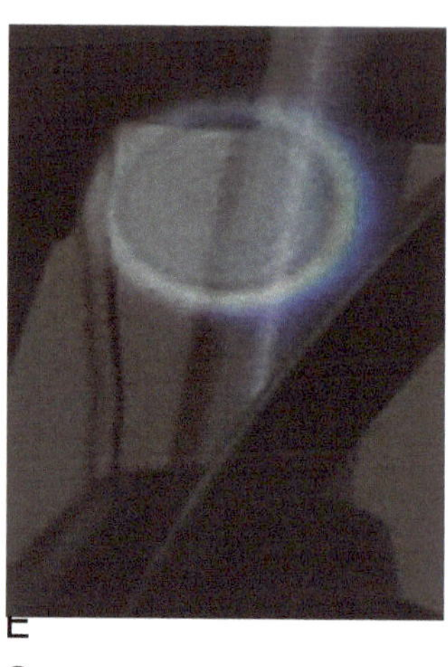

S

Second photograph with the deepening progression of the wormhole into a visible funnel.

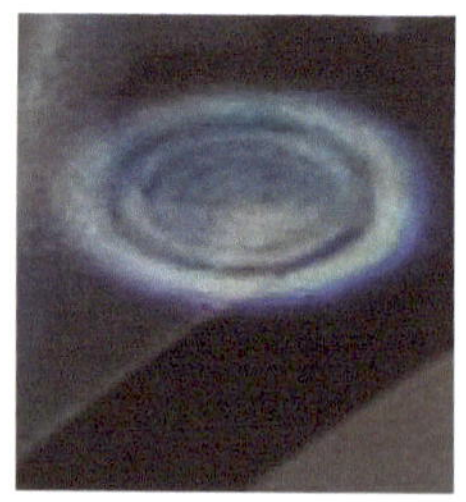

I took the photograph, below, in my family room the summer of 2013. I noticed my parrot visibly tracking energies in the room I could not see.

My attention was drawn to my laptop sitting on the kitchen table and I took a photo. In the photo you will see a magnificent, fully open stargate and a 'Being of Light' present at its entrance.

In the cropped/zoomed photo to the right, the 'Being' at the stargate looks like the mythical "Yeti".

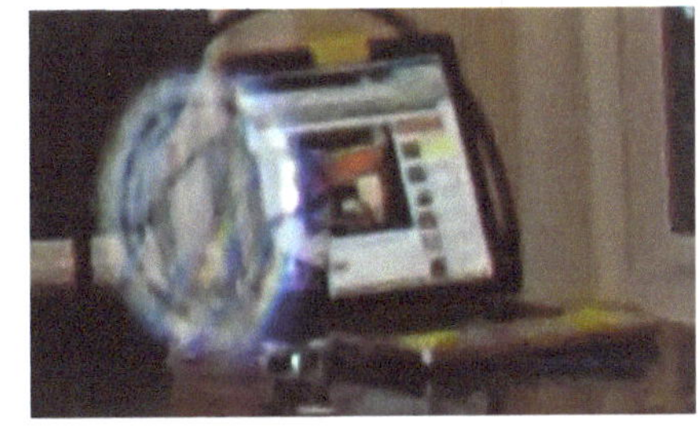

The Yeti is believed by Tibetians to be a mystical guardian of Shambhala called the "Pure Land". Like most people, I can use a vast amount of 'mental energy' at my computer. This stargate with the "Yeti" appeared at a time when I was researching subject matter to help me understand the experience I was having.

Additional interesting facts on the Yeti mythology, the pre-Buddhist shamanic mystics in the ancient Tibetan Bön religion called the Yeti "Mi rgod", and described it as a fury giant that makes a whistling sound. The Mi rgod's were allegedly guardians of the hidden city "Olmolungring", where great libraries that hold ancient wisdom reside in a pure land known as Shambhala.[45]

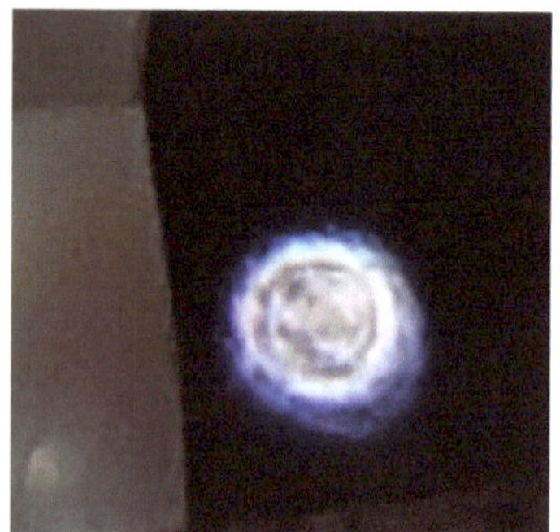

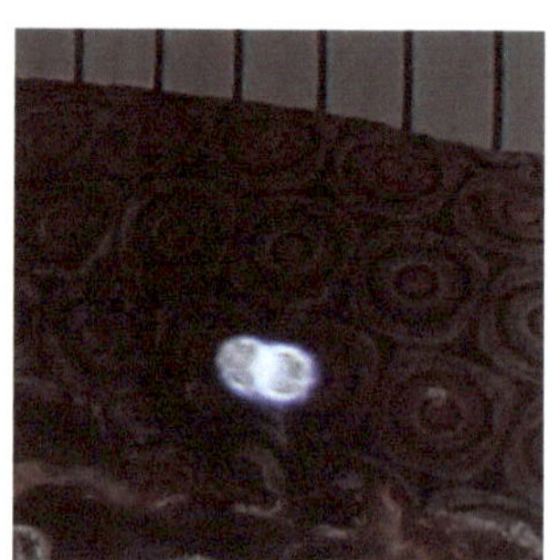

I took the above photo in my bedroom April 2014. I noticed my dog observing some 'unseen' energy near to him. I quickly picked up my camera and took this incredible photo. As you can see, in the photo are two powerful examples of photonic consciousness on opposite sides of my dog.

[45] http://www.cagliostro.se/2011/06/04/the-yeti-3071831

The sphere to the left resembles the atom and cosmic spiral, and the sphere to the right resembles the vesica pisces. Below I've created a collage of photos and images comparing the subtle energy of photonic consciousness captured in my photograph with the Milky Way galaxy, and the atom.

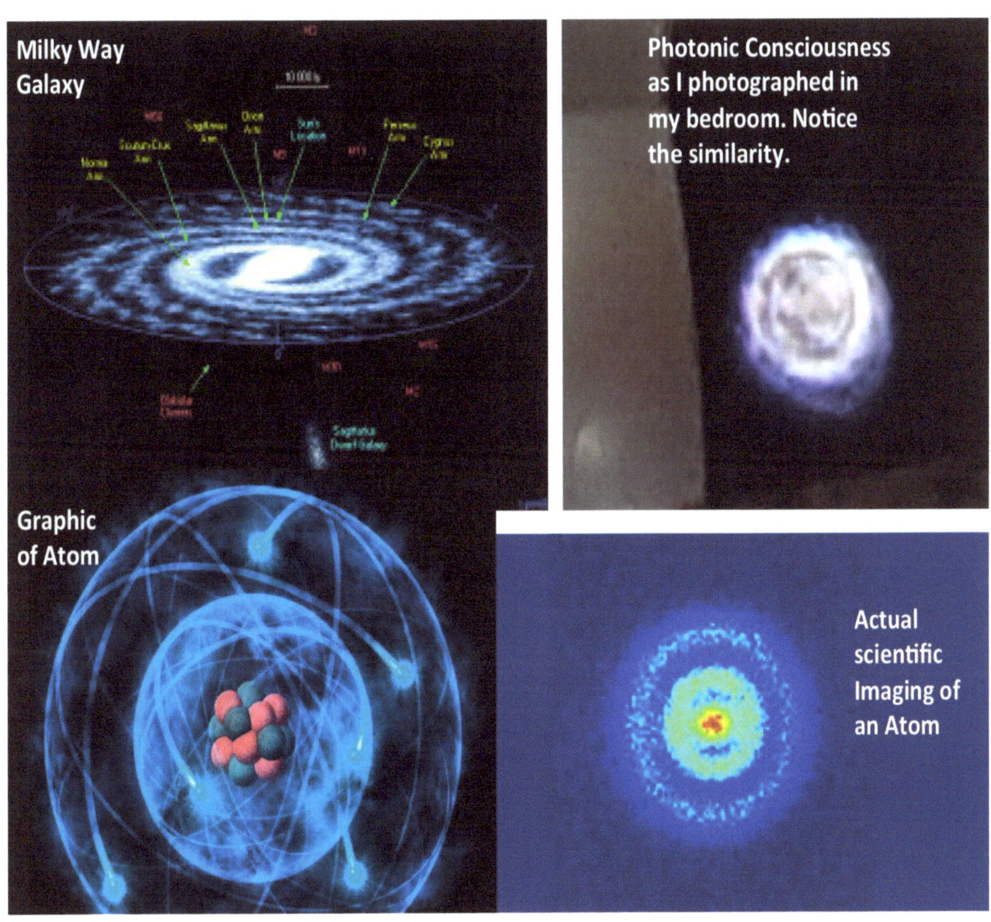

Below I've created an additional collage of photos and images comparing the subtle energy of photonic consciousness captured in my photograph, with the Hour Glass Nebula on the top right, a cosmic Vesica Pisces, and a painting from a Church depicting Jesus inside the Vesica Pisces sacred geometry on the bottom left. On the bottom right is my photograph.

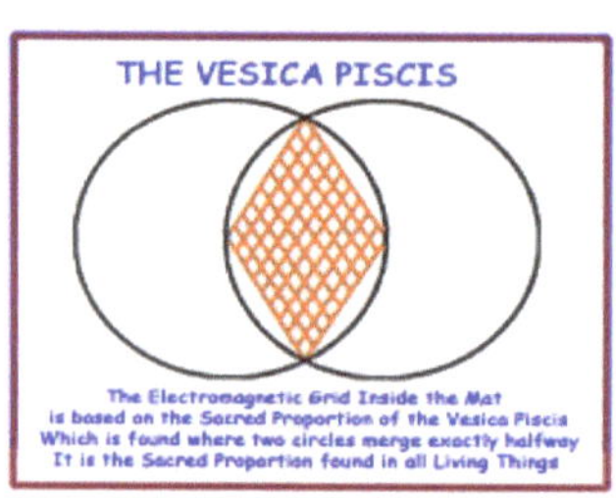

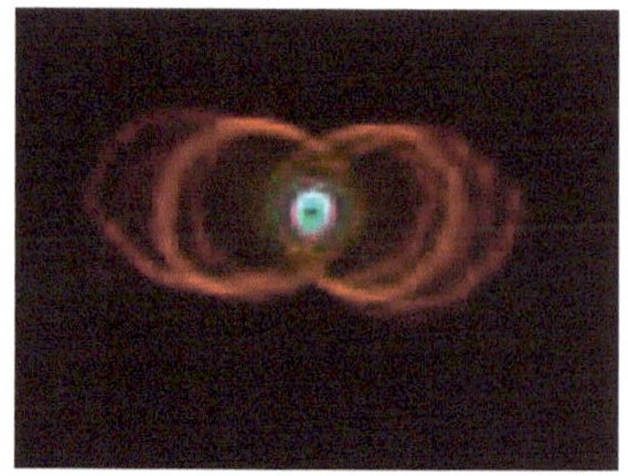

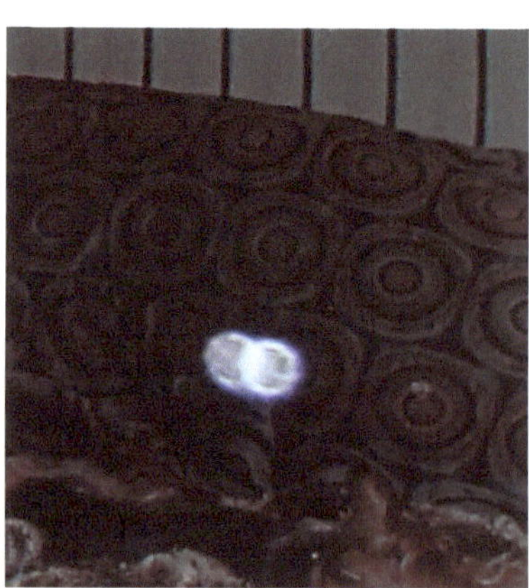

As a Christian I feel compelled to ask the fundamental question; why doesn't the Vatican share this cosmic truth with the world? The images of Jesus in churches clearly depict the geometry of creation and the phenomenon of spheres, in which Angels are depicted as residing inside. From my perspective this seems to illustrate the sphere, in this artistic rendering, as a higher dimensional 'vehicle' by which Angels travel through dimensions. Maybe one day the Vatican will explain their understanding of this phenomenon and share any evidence they may have with the rest of the world.

Lastly, I want to share with you a photo I took in April 2014 while in Egypt staying with a friend who lives close to the Great Pyramid of Giza. From my friend's rooftop you can see the Great Pyramid. The morning before my departure I went to the roof to meditate at sunrise. I just finished my meditation as the sun began to rise and took this photo. In the photo you will see an incredible sphere of light next to the top of the Great Pyramid.

The Great Pyramid is known to be a catalyst for energy capable of opening stargates. Pyramids have long been associated as being used as stargates. Pyramids can focus Earth's vibration energies towards its apex. People studying the pyramid phenomenon suggest that pyramids act like antenna's collecting electromagnetic frequencies from Earth's ionosphere. I believe that pyramids, such as the Great Pyramid of Giza, can focus and concentrate energy and these structures are powerful tools for initiation into higher consciousness; as well as maybe having once been involved in the production of healing manna, a mono atomic gold substance used to aid the body in its electrical flow and the absorption of more light. For more information on manna and the Great Pyramid connection I suggest you read Spencer L. Cross's book, *The Great Pyramid: A Factory for Monoatomic Gold*.

CHAPTER NINE

SUPER CONSCIOUSNESS

"When Christ Consciousness descends into the Soul and pure mind of man, it is called Super Consciousness."
Paramahansa Yogananda

Consciousness can be understood as an awareness of one's own state of being, existence, feelings, thoughts and surroundings. It is far greater than a state of wakefulness and cannot be reduced to a physiological response to cerebral activity. Consciousness holds infinite degrees of awareness and is multidimensional; crossing all boundaries held within the space-time continuum.

To perceive your physical world as the only reality is limiting your innate "right" to bask in the higher frequencies of light and fill your vessel (mind, body and Soul) with the infinite knowing held within the quantum information field of creation.

Our innate right as quantum, eternal Beings of Light is to proclaim our cosmic "Declaration of Independence";

> *"We hold these truths to be self-evident, that all men [and women] are created equal, that they are endowed by their Creator with certain unalienable Rights, that among these are Life, Liberty and the pursuit of Happiness..."*[46]

And, I would add, the right to truth, wisdom, and love held within God's cosmic consciousness. By our very creation, we imbue the essence of expansive quantum information encoded within our DNA – the frozen photonic story and song of humanity and all life on Earth.

All we must do to reclaim this knowledge is "know" that it is our birthright. "Knowing" (commonly thought of as "belief") is a powerful force that gives rise to the reality around us and ignites our super consciousness. To "know" is to bring it into your reality the energy of the super consciousness and in doing so we accelerate our awakening enabling us to "stand in our power" as infinite Beings of Light.

The Super Consciousness is the quantum information field of infinite knowing and wisdom. It is accessed by expanding our consciousness through awareness of self and releasing ourself from dogmatic ideas and belief systems that limit our "knowing".

This higher source of knowing, your innate super consciousness, can be felt as new ideas, inspirations, breakthroughs in understanding, desires to redirect your life, needing to let go and move forward, feelings of expansive love and joy, and the intuitive knowing that something, an idea or someone, signifies a truth. Our super consciousness 'channel' to the divine is innate. All we must do is to "know", and allow this connection to be felt within.

In this, the "I AM" presence and positive declarations of self are affirmations of this great truth. To say to yourself, "I AM the light", "I AM love", "I AM happy", "I AM abundance", "I AM truth", "I AM eternal" etc... You declare to the quantum

[46] http://www.archives.gov/exhibits/charters/declaration_transcript.html

information field and the cosmic conscious of God that you are ready to accept the responsibility and transitions of self that come with your super consciousness awakening. The positive "I AM" presence is an affirmation to self, and God, of love, respect, gratitude and grace.

I challenge you to say a positive, affirmative statement to yourself using the "I AM" presence three times every day for 30 days, and believe in its power and purpose. I promise in doing so you will feel love, joy and more expansiveness, which will empower your co-creative capabilities as you resonate in harmony and frequency with Cosmic Consciousness.

Accessing Your Super Consciousness

The I AM conscious intention, and meditation are the key ingredients enabling us to unlock our innate power. Meditation is a mental exercise that benefits the body, mind and mental processes. It is awareness of "self", quieting the mind, and stilling the body through conscious intention.

In my book *Approaching Singularity: The Genesis of Crea*tion I write about the charka energy centers in the body where photonic consciousness is metabolized and used by the body for physical, mental and emotional well being. The chakras collaborate with the body, mind and Soul, the Earth, cosmos and higher dimensional realms of existence, to manifest your experiences in life, and activate your super consciousness.

As you consciously intend to bring into your body, mind and Soul the love and light of creation, you begin to attract the higher frequencies of photonic consciousness to you and through absorption of more quantum information (light) you ascend in frequency. This is the process of invoking the Christ Consciousness, and activating your super consciousness co-creative capabilities.

We, as infinite Beings of Light, are prewired for this process. Along our spine are energy centers, known as chakras, which are the store houses metabolizing energy. We can attract negative energy into our body and space, as easily as we can attract positive, loving energy. It is all a matter of conscious intention and what frequency do we "tune our dial" too.

The chakras are a magnificent system of energy centers that relate to different levels of consciousness, developmental stages of life, emotions, thoughts, colors, sounds, body functions and much more. Bringing your chakras into balance with positive intention, will ignite your super consciousness and enable the subtle energy of photonic consciousness to awaken "you" into an enlightened state of being.

Let Your Energy Flow

As you move towards realizing the innate power of your super consciousness it is important to let your energy flow. Stagnation of the energy flowing in your body, and around your electromagnetic field, manifests as emotional, mental and physical problems with negative aspects of self.

When you balance your chakras and let you energy flow, you are in harmony with the cosmic forces of creation, Earth and the cosmos, bringing joy, love and abundance into your life.

As you master this process it exponentially increases illuminating more profound "knowings", joy and experiences. You will soon find that your consciousness is swimming in an infinite sea of potentiality and wisdom, and you can simple collect the fruits of wisdom that reside along God's lazy river of the quantum information field.

Photonic consciousness will rush in to support the manifestation of the experience(s) you desire, and enable the expansiveness of your super consciousness. This is the greatest and most profound secret of creation. It is to awaken the "super hero" from within and forever walk in the Light of God.

Message from the Author

I want to thank you for taking the time to read this book, and encourage you to explore the subjects I've presented. A pearl of wisdom I have learned in my journey is to use discernment in ascertaining truth, and to listen to my heart.

If you have enjoyed reading this book please share it with friends and family. I have free PDF downloads of all my books available at www.smokeandmirrorsbook.com. Please feel free to distribute the PDF.

May God bless your journey in life, your consciousness awakening and return to the Treasury of Light.

APPENDIX

The below pictures capture paintings in Churches, around the world, which capture the profound symbolism of Photonic Consciousness depicting Angels, Jesus and Apostles in spheres, or the Vesica Pisces. Ancient Astronaut theorists will claim this is an indication of the "Ancient Alien Theory", however I believe it signifies the Divine, multidimensional subtle energy of Photonic Consciousness. I will let you, the reader, decide.

ПРЄѠБРА
ЖЕНИЄ

I wanted to provide additional photographs I've taken over last two years. These photos have been taken in my home, and around the world. The first photo, below, was taken in an ancient rock Church in Ethiopia. The next photo was taken in Hagia Sophia in Istanbul, Turkey.

PHOTO TAKEN IN GARDEN OF GESTHEMANE, JERUSALEM 2013

PHOTO TAKEN IN THE CHURCH OF THE HOLY SEPLUCHRE, JERUSALEM 2013

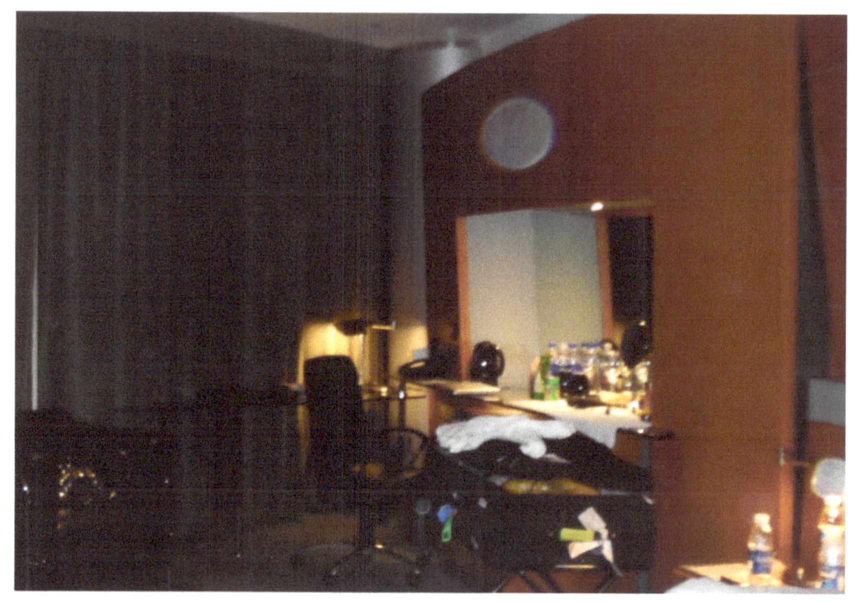

PHOTO TAKEN IN MY HOTEL ROOM IN DUBAI, U.A.E.

PHOTO TAKEN IN MY HOTEL ROOM IN TURKEY

THE BELOW PHOTOS WERE TAKEN IN MY HOME 2012-2014

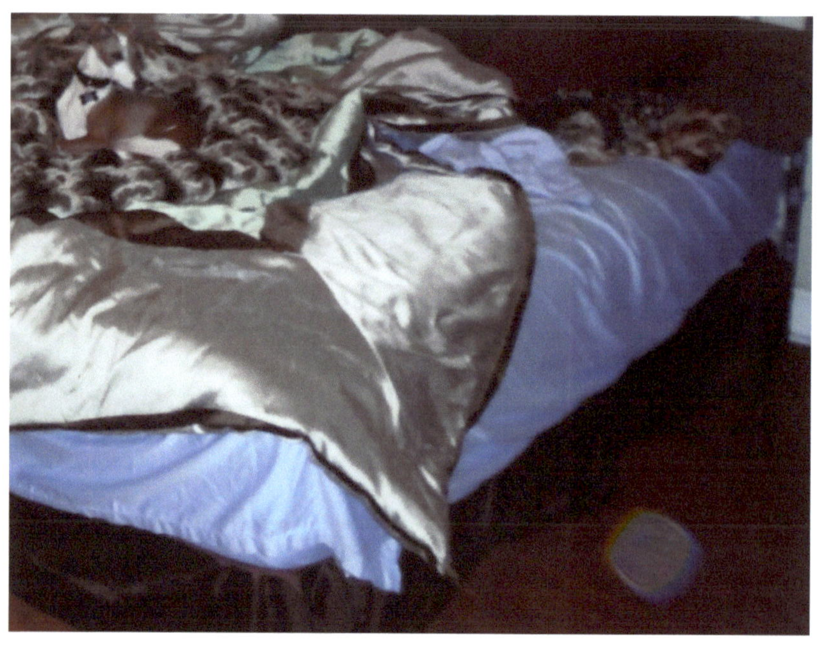

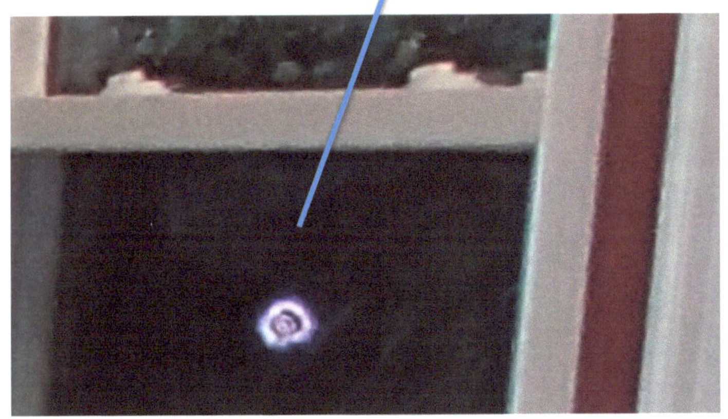

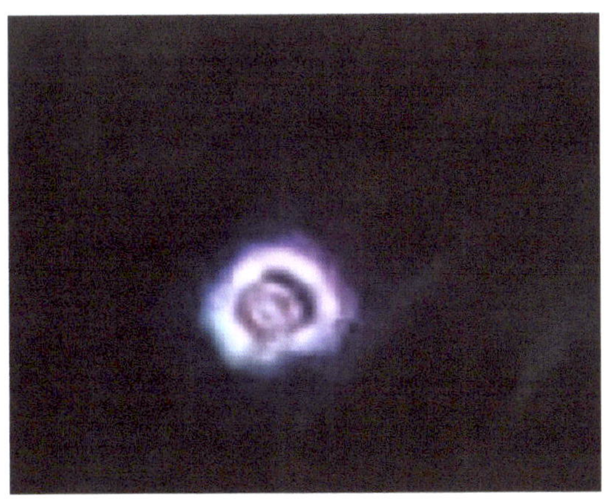

CLOSE UP OF SPHERE AND FACE INSIDE

GLOSSARY

ATOM – THE BASIC UNIT OF ORDINARY MATTER, CONSISTING OF A NUCLEUS WITH PROTONS AND NEUTRONS, SURROUNDED BY ORBITING ATOMS.

BIOPHOTON – IS A PHOTON OF NON-THERMAL ORIGIN IN THE VISIBLE AND ULTRAVIOLET SPECTRUM EMITTED FROM A BIOLOGICAL SYSTEM.

BLACK HOLE – A REGION OF SPACE-TIME THAT, DUE TO ITS IMMENSE GRAVITATION FORCE, IS CUT OFF FROM THE REST OF THE UNIVERSE.

CYMATICS - DERIVES FROM THE GREEK 'KUMA' MEANING 'BILLOW' OR 'WAVE,' TO DESCRIBE THE PERIODIC EFFECTS THAT SOUND AND VIBRATION HAVE ON MATTER.

DNA – DEOXYRIBONUCLEIC ACID, A SELF-REPLICATING MATERIAL PRESENT IN NEARLY ALL LIVING ORGANISMS AS THE MAIN CONSTITUENT OF CHROMOSOMES. IT IS THE CARRIER OF GENETIC INFORMATION.

ELECTROMAGNETIC FORCE – THE SECOND STRONGEST OF THE FOUR FORCES OF NATURE. IT ACTS BETWEEN PARTICLES WITH ELECTRIC CHARGES.

ELECTRON – AN ELEMENTARY PARTICLE OF MATTER THAT HAS A NEGATIVE CHARGE AND IS RESPONSIBLE FOR THE CHEMICAL PROPERTIES OF ELEMENTS.

EXOTIC MATTER - IN PHYSICS, EXOTIC MATTER IS A TERM WHICH REFERS TO MATTER WHICH WOULD SOMEHOW DEVIATE FROM THE NORM AND HAVE "EXOTIC" PROPERTIES.

FREQUENCY (PHYSICS) - THE RATE AT WHICH A VIBRATION OCCURS THAT CONSTITUTES A WAVE, EITHER IN A MATERIAL (AS IN SOUND WAVES), OR IN AN ELECTROMAGNETIC FIELD (AS IN RADIO WAVES AND LIGHT), USUALLY MEASURED PER SECOND.

FIBINNOCI SEQUENCE - A SERIES OF NUMBERS IN WHICH

EACH NUMBER (*FIBONACCI NUMBER*) IS THE SUM OF THE TWO PRECEDING NUMBERS. THE SIMPLEST IS THE SERIES 1, 1, 2, 3, 5, 8, ETC.

GALAXY – A LARGE SYSTEM OF STARS, INTERSTELLAR MATTER, AND DARK MATTER THAT IS HELD TOGETHER BY GRAVITY.

GOLDEN AGE - THE PERIOD WHEN A SPECIFIED ART, SKILL, ACTIVITY OR CONSCIOUSNESS IS AT ITS PEAK.

GOLDEN RATIO - ALSO KNOWN AS THE GOLDEN MEAN, OR DIVINE PROPORTION, IN MATHEMATICS, THE IRRATIONAL NUMBER $(1 + \sqrt{5})/2$, OFTEN DENOTED BY THE GREEK LETTERS T OR ϕ, AND APPROXIMATELY EQUAL TO 1.618.

GRAVITY – THE WEAKEST OF THE FOUR FORCES OF NATURE. IT IS THE MEANS BY WHICH OBJECTS THAT HAVE MASS ATTRACT EACH OTHER.

HARMONICS (PHYSICS) – A COMPONENT FREQUENCY OF AN OSCILLATION OR WAVE.

HARMONICS (MUSIC) - AN OVERTONE ACCOMPANYING A FUNDAMENTAL TONE AT A FIXED INTERVAL, PRODUCED BY VIBRATION OF A STRING, COLUMN OF AIR, ETC., IN AN EXACT FRACTION OF ITS LENGTH.

MILKY WAY GALAXY - THE GALAXY CONTAINING THE SOLAR SYSTEM; CONSISTS OF MILLIONS OF STARS THAT CAN BE SEEN AS A DIFFUSE BAND OF LIGHT STRETCHING ACROSS THE NIGHT SKY.

MULTIDIMENSIONAL - OF OR INVOLVING SEVERAL DIMENSIONS OR ASPECTS.

NASA - THE NATIONAL AERONAUTICS AND SPACE ADMINISTRATION (NASA) IS THE AGENCY OF THE UNITED STATES GOVERNMENT THAT IS RESPONSIBLE FOR THE NATION'S CIVILIAN SPACE PROGRAM AND FOR AERONAUTICS AND AEROSPACE RESEARCH.

NEGATIVE MASS - IN THEORETICAL PHYSICS, NEGATIVE MASS IS A HYPOTHETICAL CONCEPT OF MATTER WHOSE MASS IS OF OPPOSITE SIGN TO THE MASS OF NORMAL MATTER, E.G. −2 KG.

NEUTRON – A TYPE OF ELECTRICALLY NEUTRAL BARYON THAT WITH THE PROTON FORMS THE NUCLEUS OF AN ATOM.

PHASE – A POSITION IN THE CYCLE OF A WAVE.

PHOTON – A BOSON THAT CARRIES THE ELECTROMAGNETIC FORCE. A QUANTUM PARTICLE OF LIGHT.

PHOTONIC CONSCIOUSNESS – A THEORY PROPOSED IN THIS BOOK THAT EXPLAINS THE CONSCIOUS, CREATIVE FORCE OF LIGHT COMING FROM GOD.

PHOTOSYNTHESIS - THE PROCESS BY WHICH GREEN PLANTS AND SOME OTHER ORGANISMS USE SUNLIGHT TO SYNTHESIZE FOODS FROM CARBON DIOXIDE AND WATER.

PROTON – A TYPE OF POSITIVELY CHARGED BARYON THAT WITH THE NEUTRON FORMS THE NUCLEUS OF AN ATOM.

PRECESSION OF THE EQUINOXES - THE SLOW RETROGRADE, OR WESTWARD, MOTION OF EQUINOCTIAL POINTS ALONG THE ECLIPTIC.

QUANTUM THEORY – A THEORY IN WHICH OBJECTS DO NOT HAVE A SINGLE DEFINITIVE HISTORY.

QUARK – AN ELEMENTARY PARTICLE WITH A FRACTIONAL ELECTRICAL CHARGE THAT FEELS STRONG FORCE. PROTONS AND NEUTRONS ARE EACH COMPOSED OF THREE QUARKS.

RESONANCE (PHYSICS) - THE REINFORCEMENT OR PROLONGATION OF SOUND BY REFLECTION FROM A SURFACE OR BY THE SYNCHRONOUS VIBRATION OF A NEIGHBORING OBJECT.

SCHUMANN RESONANCE - A SET OF SPECTRUM PEAKS OF THE EARTH'S ELECTROMAGNETIC FIELD SPECTRUM. SCHUMANN RESONANCES ARE GLOBAL ELECTROMAGNETIC RESONANCES, EXCITED BY LIGHTNING DISCHARGES IN THE CAVITY FORMED BY THE EARTH'S SURFACE AND THE IONOSPHERE.

SPECTRUM OF LIGHT - AN ENTIRE RANGE OF LIGHT WAVES, RADIO WAVES, ETC. ACCORDING TO WAVELENGTH (OR FREQUENCY) OF ELECTROMAGNETIC RADIATION.

STARGATE – IS A PORTAL DEVICE THAT ALLOWS PRACTICAL, RAPID TRAVEL BETWEEN TWO DISTANT LOCATIONS.

STANDING WAVES – A VIBRATION OF A SYSTEM IN WHICH SOME PARTICULAR POINTS REMAIN FIXED WHILE OTHERS BETWEEN THEM VIBRATE WITH THE MAXIMUM AMPLITUDE.

SINGULARITY – A POINT IN SPACE-TIME AT WHICH A PHYSICAL QUANTITY BECOMES INFINITE.

SPACE-TIME – A MATHEMATICAL SPACE WHOSE POINTS MUST BE SPECIFIED BY BOTH SPACE AND TIME COORDINATES.

STRING THEORY – A THEORY OF PHYSICS IN WHICH PARTICLES ARE DESCRIBED AS PATTERNS OF VIBRATION THAT HAVE LENGTH BUT NO HEIGHT OR WIDTH – LIKE INFINITELY

THIN PIECES OF STRING.

SUBTLE ENERGY – ENERGY THAT CANNOT BE ACCURATELY MEASURED USING CURRENT SCIENTIFIC METHODS.

SUPER CONSCIOUSNESS – ALSO CALLED HIGHER CONSCIOUSNESS AND BUDDHIC CONSCIOUSNESS ARE EXPRESSIONS USED IN VARIOUS SPIRITUAL AND INTELLECTUAL TRADITIONS TO DENOTE THE EXPANDED CONSCIOUSNESS THAT RESIDES WITHIN THE CHRIST AND COSMIC CONSCIOUSNESS.

VESICA PISCES - IS A SHAPE THAT IS THE INTERSECTION OF TWO CIRCLES WITH THE SAME RADIUS, INTERSECTING IN SUCH A WAY THAT THE CENTER OF EACH CIRCLE LIES ON THE PERIMETER OF THE OTHER.

WORMHOLE (PHYSICS) - A HYPOTHETICAL CONNECTION BETWEEN WIDELY SEPARATED REGIONS OF SPACE-TIME.

VIBRATION - AN OSCILLATION OF THE PARTS OF A FLUID OR AN ELASTIC SOLID WHOSE EQUILIBRIUM HAS BEEN DISTURBED, OR OF AN ELECTROMAGNETIC WAVE.

BIBLIOGRAPHY

Ammon-Wexler, Dr. Jill. 2011. *Pineal Gland & Your Third Eye: Develop "Conscious Self" Psychic Abilities*.

Anonymous. Date unknown. *The Hermetic Arcanum: The Secret Work of the Hermetic Philosophy*.

Blavatsky, H.P. 1877. *Isis Unveiled*

Branden, Gregg. 2009. *Fractal Time*

Clow, Barbara Hand. 2004. *Alchemy of Nine Dimensions: The 2011/2012 Prophecies and Nine Dimensions of Consciousness*.

Dale, Cyndi. 2009. *The Subtle Body: An Encyclopedia of Your Energetic Anatomy*

Faust, Michael. 2010 *Kabbalah, Hermeticism and the M-Theory*.

Godwin, Joscelyn. 2011. *Atlantis and the Cycles of Time: Prophecies, Traditions and Occult Revelations*

Hall, Manly P. 1928. *The Secret Teachings of All Ages*

Hawking, Stephen & Mlodinow, Leonard. 2010. *The Grand Design*

King James Bible. 1769

Kreisbert, Glenn. 2010. *Lost Knowledge of the Ancients*

Martinez, Dr. Susan B. 2011. *Time of the Quickening: Prophecies for the Coming Utopian Age*

Mead, G.R.S. *1992 (1906). Thrice Greatest Hermes: Studies in Hellenistic Theosophy and Gnosis";* York Beach, Maine: Samuel Weiser

Melchizedek, Drunvalo. 2007. *Serpent of Light: The Movement of the Earth's Kundalini and the Rise of the Female Light 1949 to 2013*

Morrell, Peter. 1988. *The Path of Non-Attachment; The Dalai Lama at Harvard*

Naga, Jim. 2010. *How to Open Your 3rd Eye*

Notable quotes & quotations: http://quotes.liberty-tree.ca/quote_blog/Nathan.Mayer.Rothschild.Quote.4F94

Paracelsus, Philippus Aureolus Theophrastus Bombastus von Hohenheim. 1493. *The Book of the Revelations of Hermes: Concerning the Supreme Secret of the World*

Phylos, Orpheus, 1999. *Earth, the Cosmos and You*

Ra, Summum Bonum Amen. 1975. *Summum: Sealed Except to the Open Mind*

Ramacharacka, Yogi. 1908. *The Collective Works*

Robinson, James M. ed. 1990. *The Sophia of Jesus;* The Nag Hammadi Library, revised edition.

Roberts, J.M. 1892. *Antiquity Unveiled*

Rothman, Tony. 1995. *Instant Physics: From Aristotle to Einstine and Beyond*

Sams, Gregory. 2009. *Sun of God: Discover the Self-Organizing Consciousness that Underlies Everything*

Thoth, the Atlantean. 1993. *The Emerald Tablets of Thoth-the-Atlantean*"; translations and intrepration by Doreal

Three Initiates. 1908. *The Kybalion: A Study of the Hermetic Philosophy of Ancient Egypt and Greece*

Tresmigistus, Hermes. 1957. *The Emerald Tablet of Hermes; History of the Tablet*, Holmyard 1957 and Needham 1980